Pedestrian Inertial Navigation with Self-Contained Aiding

Pedestrian Inertial Navigation with Self-Contained Aiding

Yusheng Wang and Andrei M. Shkel
University of California, Irvine

IEEE Press Series on Sensors
Vladimir Lumelsky, Series Editor

IEEE PRESS

WILEY

Published by John Wiley & Sons, Inc., Hoboken, New Jersey.
Published simultaneously in Canada.

For general information on our other products and services or for technical support, please contact our Customer Care Department within the United States at (800) 762-2974, outside the United States at (317) 572-3993 or fax (317) 572-4002.

Wiley also publishes its books in a variety of electronic formats. Some content that appears in print may not be available in electronic formats. For more information about Wiley products, visit our web site at www.wiley.com.

Library of Congress Cataloging-in-Publication Data applied for:

ISBN: 9781119699552

Cover Design: Wiley
Cover Image: © Production Perig/Shutterstock

Set in 9.5/12.5pt STIXTwoText by Straive, Chennai, India

10 9 8 7 6 5 4 3 2 1

Contents

Author Biographies

Yusheng Wang, PhD, received the B.Eng. degree (Hons.) in engineering mechanics from Tsinghua University, Beijing, China, in 2014 and the Ph.D. degree in mechanical and aerospace engineering from the University of California, Irvine, CA, in 2020. His research interests include the development of silicon-based and fused quartz-based MEMS resonators and gyroscopes, and pedestrian inertial navigation development with sensor fusion. He is currently working at SiTime Corporation as an MEMS Development Engineer.

Andrei M. Shkel, PhD, has been on faculty at the University of California, Irvine since 2000, and served as a Program Manager in the Microsystems Technology Office of DARPA. His research interests are reflected in over 300 publications, 42 patents, and 3 books. Dr. Shkel has been on a number of editorial boards, including Editor of *IEEE/ASME JMEMS, Journal of Gyroscopy and Navigation*, and the founding chair of the *IEEE Inertial Sensors*. He was awarded the Office of the Secretary of Defense Medal for Exceptional Public Service in 2013, and the 2009 IEEE Sensors Council Technical Achievement Award. He is the President of the IEEE Sensors Council and the IEEE Fellow.

List of Figures

List of Tables

1

Introduction

1.1 Navigation

Navigation is the process of planning, recording, and controlling the movement of a craft or vehicle from one place to another [1]. It is an ancient subject but also a complex science, and a variety of methods have been developed for different circumstances, such as land navigation, marine navigation, aeronautic navigation, and space navigation.

One of the most straightforward methods is to use landmarks. Generally speaking, a landmark can be anything with known coordinates in a reference frame. For example, any position on the surface of the Earth can be described by its latitude and longitude, defined by the Earth's equator and Greenwich meridian. The landmarks can be hills and rivers in the wilderness, or streets and buildings in urban areas, or lighthouses and even celestial bodies when navigating on the sea. Other modern options, such as radar stations, satellites, and cellular towers, can all be utilized as landmarks. The position of the navigator can be extracted by measuring the distance to, and/or the orientation with respect to the landmarks. For example, celestial navigation is a well-established technique for navigation on the sea. In this technique, "sights," or angular distance is measured between a celestial body, such as the Sun, the Moon, or the Polaris, and the horizon. The measurement, combined with the knowledge of the motion of the Earth, and time of measurement, is able to define both the latitude and longitude of the navigator [2]. In the case of satellite navigation, a satellite constellation composed of many satellites with synchronized clocks and known positions, and continuously transmitting radio signal is needed. The receiver can measure the distance between itself and the satellites by comparing the time difference between the signal that is transmitted by the satellite and received by the receiver. A minimum of four satellites must be in view of the receiver for it to compute the time and its location [3]. Navigation methods of this type, which utilize the observation of

Pedestrian Inertial Navigation with Self-Contained Aiding, First Edition. Yusheng Wang and Andrei M. Shkel.
© 2021 The Institute of Electrical and Electronics Engineers, Inc. Published 2021 by John Wiley & Sons, Inc.

landmarks with known positions to directly determine a position, are called the position fixing. In the position fixing type of navigation, navigation accuracy is dependent only on the accuracy of the measurement and the "map" (knowledge of the landmarks). Therefore, navigation accuracy remains at a constant level as navigation time increases, as long as observations of the landmarks are available.

The idea of position fixing is straightforward, but the disadvantage is also obvious. Observation of landmarks may not always be available and is susceptible to interference and jamming. For example, no celestial measurement is available in foggy or cloudy weather; radio signals suffer from diffraction, refraction, and Non-Line-Of-Sight (NLOS) transmission; satellite signals may also be jammed or spoofed. Besides, a known "map" is required, which makes this type of navigation infeasible in the completely unknown environment.

An alternative navigation type is called dead reckoning. The phrase "dead reckoning" probably dated from the seventeenth century, when the sailors calculated their location on the sea based on the velocity and its orientation. Nowadays, dead reckoning refers to the process where the current state (position, velocity, and orientation) of the system is calculated based on the knowledge of its initial state and measurement of speed and heading [4]. Velocity is decomposed into three orthogonal directions based on heading and then multiplied by the elapsed time to obtain the position change. Then, the current position is calculated by summing up the position change and the initial position. A major advantage of dead reckoning over position fixing is that it does not require the observations of the landmarks. Thus, the system is less susceptible to environmental interruptions. On the other hand, dead reckoning is subject to cumulative errors. For example, in automotive navigation, the odometer calculates the traveled distance by counting the number of rotations of a wheel. However, slipping of the wheel or a flat tire will result in a difference between the assumed and actual travel distance, and the error will accumulate but cannot be measured or compensated, if no additional information is provided. As a result, navigation error will be accumulated as navigation time increases.

Inertial navigation is a widely used dead reckoning method, where inertial sensors (accelerometers and gyroscopes) are implemented to achieve navigation purpose in the inertial frame. The major advantage of inertial navigation is that it is based on the Newton's laws of motion and imposes no extra assumptions on the system. As a result, inertial navigation is impervious to interference and jamming, and its application is universal in almost all navigation scenarios [5].

1.2 Inertial Navigation

The operation of inertial navigation relies on the measurements of accelerations and angular rates, which can be achieved by accelerometers and gyroscopes, respectively. In a typical Inertial Measurement Unit (IMU), there are three accelerometers and three gyroscopes mounted orthogonal to each other to measure the acceleration and angular rate components along three perpendicular directions. To keep track of the orientation of the system with respect to the inertial frame, three gyroscopes are needed. Gyroscopes measure the angular rates along three orthogonal directions. Angular rates are then integrated, and the orientation of the system is derived from these measurements. The readout of the accelerometers is called the specific force, which is composed of two parts: the gravity vector and the acceleration vector. According to the Equivalence Principle in the General Theory of Relativity, the inertial force and the gravitational force are equivalent and cannot be separated by the accelerometers. Therefore, the orientation information obtained by the gyroscopes is needed to estimate the gravity vector. With the orientation information, we can subtract the gravity vector from the specific force to obtain the acceleration vector, and revolve the acceleration vector from the system frame to the inertial frame before performing integration. Given the accelerations of the system, the change of position can be calculated by performing two consecutive integrations of the acceleration with respect to time.

The earliest concept of inertial sensor was proposed by Bohnenberger in the early nineteenth century [6]. Then in 1856, the famous Foucault pendulum experiment was demonstrated as the first rate-integrating gyroscope [7], whose output is proportional to the change of angle, instead of the angular rate as in the case of most commercial gyroscopes. However, the first implementation of an inertial navigation system did not occur until the 1930s on V2 rockets and the wide application of inertial navigation started in the late 1960s [8]. In the early implementation of inertial navigation, inertial sensors are fixed on a stabilized platform supported by a gimbal set with rotary joints allowing rotation in three dimensions (Figure 1.1).

Figure 1.1 A schematic of gimbal system. Source: Woodman [5].

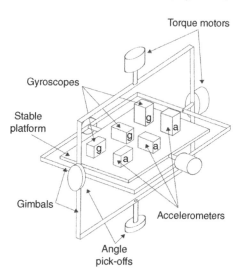

The gyroscope readouts are fed back to torque motors that rotates the gimbals so that any external rotational motion could be canceled out and the orientation of the platform does not change. This implementation is still in common use where very accurate navigation data is required and the weight and volume of the system are not of great concern, such as in submarines. However, the gimbal systems are large and expensive due to their complex mechanical and electrical infrastructure. In the late 1970s, strapdown system was made possible, where inertial sensors are rigidly fixed, or "strapped down" to the system. In this architecture, the mechanical complexity of the platform is greatly reduced at the cost of substantial increase in the computational complexity in the navigation algorithm and a higher dynamic range for gyroscopes. However, recent development of microprocessor capabilities and suitable sensors allowed such design to become reality. The smaller size, lighter weight, and better reliability of the system further broaden the applications of the inertial navigation. Comparison of the schematics of algorithmic implementations in gimbal system and strapdown system is shown in Figure 1.2.

Inertial navigation, as a dead reckoning approach to navigation, also suffers from error accumulations. In the inertial navigation algorithm, not only accelerations and angular rates are integrated but all the measurement noises are also integrated and accumulated. As a result, unlike the position fixing type of navigation, the navigation accuracy deteriorates as navigation time increases. Noise sources include fabrication imperfections of individual inertial sensors, assembly errors of the entire IMU, electronic noises, environment-related errors (temperature, shock, vibration, etc.), and numerical errors. Thus, inertial navigation imposes challenging demands on the system, in terms of the level of errors,

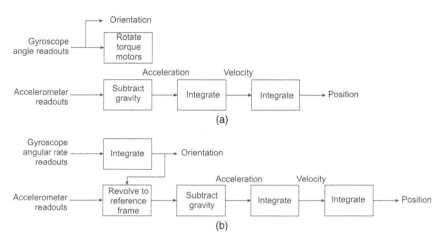

Figure 1.2 Comparison of (a) gimbal inertial navigation algorithm and (b) strapdown inertial navigation algorithm.

to achieve long-term navigation. This partially explains why inertial navigation systems were developed around 100 years later than the development of inertial sensors. It has been shown that without an error-suppressing algorithm, the position error accumulates without bound and approximately proportional to time cubed. For example, for navigation grade IMUs, which cost a few hundred thousand dollars per axis, the navigation error will reach about one nautical mile after an hour of navigation, or equivalently less than 0.01 m of navigation error within a minute of navigation. However, for consumer grade IMUs, which cost a few dollars, the navigation error will exceed a meter of error within a few seconds of navigation [9]. Therefore, aiding techniques are necessary to limit the navigation error propagation in inertial navigation, especially in the case of pedestrian inertial navigation, where the cost and size of the system are limited.

1.3   Pedestrian Inertial Navigation

Pedestrian navigation has been of great interest in recent years for path finding, personal security, health monitoring, and localizers for first responder systems. Due to the complicated environment in which a person may need to navigate, self-contained navigation techniques are fundamental for pedestrian navigation. An example of the self-contained navigation technique is inertial-only navigation of pedestrians, which became recently a popular topic. Most pedestrian navigation systems rely on inertial sensors and inertial navigation techniques in their core, just as any other navigation applications. However, the pedestrian navigation poses much stricter requirements on the size and weight of inertial instruments, or IMUs, due to the limitation of human carrying capacity, and the inertial-only pedestrian application was technologically not feasible until recently.

Thanks to the development of Micro-Electro-Mechanical Systems (MEMS) technology in the past 20 years, MEMS-based IMUs have become smaller in size and more accurate in performances, and as a result, pedestrian inertial navigation has been made possible [10]. MEMS-based IMUs with a size on the order of millimeters have become widely available on the market, and they can be installed in portable devices that can be easily carried around, such as mobile phones, smart watches, or devices that are small enough to be carried in a pocket. Figure 1.3 compares the IMU that was developed for the Apollo missions 50 years ago and a current commercial MEMS-based IMU. This is an illustration of technological advances in size, and it should be acknowledged that performances of the two systems are still not the same. Note that a gimbal inertial navigation was implemented for the Apollo mission, instead of the more commonly used strapdown inertial navigation systems in these days. The IMU for the Apollo missions had a volume of 1100 in^3 (or 1.8×10^7 mm^3) and a weight of 42.5 lb [11], whereas the volume of

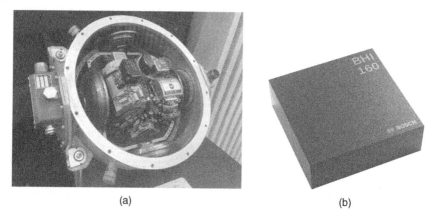

(a)   (b)

Figure 1.3 A comparison of (a) an IMU developed for the Apollo missions in 1960s. Source: https://en.wikipedia.org/wiki/Inertial_measurement_unit and (b) a current commercial MEMS-based IMU. Source: https://www.bosch-sensortec.com/products/smart-sensors/bhi160b/.

the shown MEMS-bases IMU is $8.55\,\text{mm}^3$ and the weight is on the order of tens of milligrams. Six orders of magnitude of reduction in both volume and weight has been demonstrated over the past 50 years, though to achieve the matching performance is still an on-going area of research. Such a great technical advancement in the miniaturization of IMUs started enabling the pedestrian inertial navigation. Along with the size reduction, the performance of inertial sensors is continuously improving. The use of miniaturized sensors in these new applications inspired the development of new algorithms and new approaches for solving the challenges of navigation. These approaches are discussed next.

1.3.1 Approaches

There are two general approaches in the pedestrian inertial navigation. One is the strapdown inertial navigation as introduced in Section 1.2, where IMU readouts are integrated into position and orientation. This approach is universally applicable, but the integral step makes the algorithm computationally expensive and the navigation error accumulates as time cubed due to the gyroscope bias. In order to limit the error propagation, the most commonly used method is to apply the Zero-Velocity Updates (ZUPTs) when the velocity of the foot is close to zero (the foot is stationary on the ground) [12]. The stationary state can be used to limit the long-term velocity and angular rate drift, thus greatly reduce the navigation error. In this implementation, IMU is fixed on the foot to perform the navigation and to detect the stance phase at the same time. Whenever the stance phase is detected, the zero-velocity information of the foot is fed into the Extended

Kalman Filter (EKF) as a pseudo-measurement to compensate for IMU biases, thus reducing the navigation error growth in the system. In this architecture, not only the navigation errors but also IMU errors can be estimated by the EKF. The limitation of this approach is that the IMU needs to be mounted on the foot.

In order to avoid the integral step in the pedestrian inertial navigation and also relax the requirement of IMU mounting position, a Step-and-Heading System (SHS) is an alternative. It is composed of three main parts: step detection, step length estimation, and step heading angle estimation [13]. Unlike the first approach, this approach can only be applied in the pedestrian inertial navigation. In this approach, the step length of each stride is first estimated based on some features of motion obtained from the IMU readouts. Methods based on biomechanical models and statistical regression methods are popular for the estimation. Some commonly used features include the gait frequency, magnitude of angular rate, vertical acceleration, and variance of angular rate. Then, the heading angle is estimated by the gyroscope readout, which is typically mounted at the head. This step can also be aided by magnetometers to improve the accuracy. In this way, the total displacement can be estimated combining the traveled distance and the heading angle. However, two major challenges exist for this approach. First, the gazing direction needs to be aligned with the traveling direction, implying that the subject needs to look at the traveling direction all the time, which is not practical. Second, the step length estimation remains difficult. The average value of the estimated step length may be accurate when median value generally less than 2%, but the estimate precision is generally low, with the Root Mean Square Error (RMSE) about 5% [14]. With a wide adaption of hand-held and fitness devices, this is currently an active area of research.

1.3.2 IMU Mounting Positions

In pedestrian inertial navigation, depending on the approaches to be taken and the application restrictions, the IMU can be mounted on different parts of body to take advantage of different motion patterns, such as head, pelvis, foot, wrist, thigh, and foot. Pelvis, or lower back, was the first explored IMU mounting position in the literature, because these parts of the body experience almost no change of orientation during walking, which greatly simplifies the modeling process for both strapdown inertial navigation and SHS [15]. In subsequent studies, thigh and shank were explored, such that IMU can directly measure the motion of the leg, which is directly related to the step length by the biomechanical models [16, 17]. More recently, in order to integrate pedestrian inertial navigation with smart phones and wearable devices, such as smart watches and smart glasses, pocket, wrist (or hand hold), and head are becoming the IMU mounting positions of interest [18–20]. The foot-mounted IMU has also been demonstrated for SHS, but this placement of sensors is mostly used in the ZUPT-aided pedestrian inertial navigation, instead of SHS.

Head-mounted IMUs are usually used for heading angle estimation, since it experiences lowest amount of shock and almost no change of orientation. Besides, it is usually convenient to mount the IMU on the helmet for first responders and military applications [21]. However, the low amplitude of angular rate and acceleration during walk makes it hard for step length detection. In addition, the gazing direction may not be aligned with walking direction during navigation. Pelvis-mounted IMUs have the ability of estimating the step length for both legs with one single device, compared to the IMUs mounted on the legs. It is also more convenient to align the IMU to the walking direction compared to the head-mounted IMUs. Pocket-mounted IMUs and hand-held IMUs are mostly developed for pedestrian inertial navigation for use with smart phones. In this approach, the IMU is not fixed to a certain part of the body, and its orientation may change over the navigation applications due to different hand poses and different ways to store the smart phone in the pocket. It makes the SHS algorithm more complicated than other IMU mounting positions. Foot-mounted IMUs will experience the highest amount of shock and vibration due to the heel shocks during walking [22]. As a result, a more stringent requirement on the IMU performance will be necessary, such as high shock survivability, high bandwidth and sampling rate, low g-sensitivity, and low vibration-induced noise [23]. However, with foot-mounted IMUs, a close-to-stationary state of the foot during the stance phases will greatly reduce the navigation errors in the ZUPT-aided pedestrian inertial navigation.

1.3.3 Summary

Between the SHS and the ZUPT-aided strapdown inertial navigation, the latter is the more widely used approach for precision pedestrian inertial navigation. The main reasons are:

- ZUPT-aided strapdown inertial navigation has demonstrated a better navigation accuracy compared to the SHS. For example, in a navigation with the total walking distance of 20 km, position estimation error on the order of 10 m was demonstrated, corresponding to a navigation error less than 0.1% of the total distance [24]. The navigation error for SHS, however, is typically about 1% to 2% of the total walking distance.
- ZUPT-aided strapdown inertial navigation is more universal compared to SHS, with only one assumption that the velocity of the foot is zero during the stance phase. As a result, it can be applied to many pedestrian scenarios, such as walking, running, jumping, and even crawling. In the case of SHS, it has to classify

different motion patterns, if the system has been trained with such patterns, and correspondingly fit the data to different models.

- SHS is user-specified and needs to be calibrated or trained according to different subjects, while ZUPT-aided strapdown inertial navigation in principle does not need any special calibration for different users.
- Even though IMU will experience high level of shock and vibration when mounted on the foot in the ZUPT-aided strapdown inertial navigation, the developed MEMS technologies are able to reduce the disadvantageous effects. For example, it has been demonstrated that IMU with gyroscope maximum measuring range of $800° \text{ s}^{-1}$ and bandwidth of 250 Hz would be able to capture most features of the motion without causing large errors [25].

In this book, we will mainly focus on the ZUPT-aided strapdown inertial navigation.

1.4 Aiding Techniques for Inertial Navigation

Many aiding techniques have been developed to fuse with inertial navigation to improve the navigation accuracy. They can be roughly categorized into self-contained aiding and aiding that relies on external signals (non-self-contained aiding). We start with non-self-contained aiding.

1.4.1 Non-self-contained Aiding Techniques

According to the property of the external signals, non-self-contained aiding techniques can be divided into two categories. In the case where the external signals are naturally existent, such as the Earth's magnetic field and the atmospheric pressure, no extra infrastructure is needed, but the signals may be subject to disturbance since their sources are not controlled. However, in the other case where man-made signals are used, implementation of infrastructures is needed with the benefit that the signals are engineered to facilitate the navigation process.

1.4.1.1 Aiding Techniques Based on Natural Signals

Magnetometry and barometry are two commonly applied techniques that are used to improve the navigation accuracy. Magnetometry is one of the most ancient aiding techniques developed for navigation applications, where measurement of the Earth's magnetic field can provide information about the orientation of the system. Nowadays, not just the orientation, but also the location of the system can be obtained by measuring the anomalies of the Earth's magnetic field in the navigation of low-earth-orbiting spacecraft (altitude less than 1000 km), where the position of the spacecraft can be estimated with resolution on the order of 1 km. Barometry estimates the altitude of the system by measuring the atmospheric air pressure. It has been shown that low altitudes above sea level, the atmospheric pressure decreases approximately linearly as the altitude increases with a rate of about 12 Pa/m. A pressure measurement resolution of 1 Pa, or an altitude measurement accuracy of less than 0.1 m, can be achieved with the currently available commercial micro barometers [26].

Another way of implementing estimations of absolute position is through computer vision, where images of the environment are captured to extract information. One of the most popular implementations is called the Simultaneous Localization and Mapping (SLAM), where the localization and mapping of the environment is conducted simultaneously. As a result, no pre-acquired database of the environment is needed. The sensors used for this application do not necessarily have to be cameras, LIght Detection And Ranging (LIDAR) and ultrasonic ranging can also be used. In either case, the system extracts some information about the environment as an aiding technique to improve the navigation accuracy.

1.4.1.2 Aiding Techniques Based on Artificial Signals

Radio-based navigation is another popular technique in this category. It was first developed in the early twentieth century and its application was widely developed in the World War II. More recently, it was considered as a reliable backup of the global positioning system (GPS) in the United States, and could reach a navigation accuracy of better than 50 m. One of the most common aiding techniques in this category is Global Navigation Satellite System (GNSS), where a satellite constellation is implemented in the space as "landmarks," transmitting radio waves for navigation purposes. The navigation accuracy of GNSS for civilian use is currently about 5 m along the horizontal direction, and about 7.5 m along the vertical direction. Long-term evolution (LTE) signals have also been proposed and demonstrated to be used for navigation purposes. The principle of LTE-based navigation is similar to GNSS, except that the landmarks are the LTE signal towers instead of satellites. The greatest advantage of LTE over GNSS is its low cost, since no special signal towers has to be established and maintained. Currently, a horizontal navigation accuracy of better than 10 m has been reported.

In the case of short-range navigation aiding techniques, Ultra-Wide Band (UWB) radio, WiFi, Bluetooth, and Radio-Frequency Identification (RFID) have all been explored. They are typically used in indoor navigation due to their short signal propagation range. Unlike radio-based navigation, in which the radio frequency is fixed, UWB radio occupies a large bandwidth (>500 MHz), thus increasing capability of data transmission, range estimation accuracy, and material penetration. WiFi and Bluetooth devices are popular in smartphones, and therefore utilizing them as aiding techniques in indoor navigation does not require any additional infrastructures. RFID has also been proposed due to its low cost for implementation. More recently, 5G and millimeter-wave communication infrastructure have been explored as a potential source of signals for navigation [27]. For all these aiding techniques, there are two kinds of methods to perform localization: Received Signal Strength (RSS) and fingerprinting. RSS-based localization algorithm takes advantage of the fact that the strength of the received signal drops as the distance between the source and the receiver increases. Therefore, the strength of the received signal can be used as an indicator of ranging information. Fingerprinting localization algorithm is based on comparing the measured RSS values with a reference map of RSS. Table 1.1 summarizes the non-self-contained aiding techniques with artificial signals.

1.4.2 Self-contained Aiding Techniques

Another category of aiding techniques is self-contained aiding. Instead of fusing external signals into the system, self-contained aiding takes advantage of the system's patterns of motion to compensate for navigation errors. Therefore, self-contained aiding techniques vary for different navigation applications due to different dynamics of motions.

For example, in ground vehicle navigation, the wheels can be assumed to be rolling without slipping. Thus, IMU can be mounted on the wheel of the vehicle to take advantage of the rotational motion of the wheel. In this architecture, the velocity of the vehicle can be measured by multiplying the rotation rate of the wheel by the circumference of the tire [28]. In addition, carouseling motion of the IMU provides the system more observability of the IMU errors, especially the error of yaw gyroscope, which is typically nonobservable in most navigation scenarios [29]. Besides, low frequency noise and drift can also be reduced by algorithms taking advantage of the motion of the IMU [30].

Another approach is to take advantage of biomechanical model of human gait instead of just the motion of the foot during walking. This approach typically requires multiple IMUs fixed on different parts of human body and relate the recorded motions of different parts through some known relationships derived from the biomechanical model. In this approach, a more accurate description of the human gait is available, through for example, human activity classification and human gait reconstruction. Recognition of gait pattern can help to reduce the navigation error obtained from a single IMU.

Machine Learning (ML) has also been applied to pedestrian inertial navigation. ML has mostly been explored in the field of Human Activity Recognition (HAR) [31], stride length estimation [32], and stance phase detection [33]. However, few studies used the ML approach to directly solve the pedestrian navigation problem. Commonly used techniques include Decision Trees (DT) [34], Artificial Neural

Table 1.1 Summary of non-self-contained aiding techniques.

Aiding technique	Applicable area	Positioning accuracy (m)	Notes
GPS	Above earth surface	5	Large signal coverage area
			Unavailable below the Earth's surface and in complex urban areas
			Susceptible to jamming and spoofing
LTE/5G	Mostly in urban areas	10	No extra infrastructure needed
			Rely on cellular signal coverage
Radar	In the air	50	Cheap and robust to different weathers
			Very large effective range
			Signal can penetrate insulators but will be obstructed by conductive material
UWB	Mostly indoor	0.01	Very accurate distance measurement in a short range
			Simple hardware with low power consumption
			Susceptible to interference
Lidar	In the air	0.1	Accurate position and velocity measurement
			Affected by the weather, such as strong sunlight, cloud, and rain
WiFi	Indoor	1	A priori knowledge of WiFi router is needed
			Algorithm is needed to compensate for signal strength fluctuations
Bluetooth	Indoor	0.5	Moderate measurement accuracy with very low power hardware
			Short range of measurement (<10 m)
RFID	Indoor	2	Easy deployment
			Very short range of measurement

Network (ANN) [35], Convolutional Neural Network (CNN) [36], Support Vector Machine (SVM) [37], and Long Short-Term Memory (LSTM) [38].

Sensor fusion method can be used in a self-contained way, where multiple self-contained sensors are used in a single system, and their readouts are fused in the system to obtain a navigation result. For example, self-contained ranging technique is one possibility. In this technique, the transmitter sends out a signal (can be ultrasonic wave or electromagnetic wave) which is received by the receiver. This technique can be categorized as self-contained if both the transmitter and the receiver are within the system whose state is to be estimated. For example, in the foot-to-foot ranging, the transmitter and receiver are placed on two feet of a person to keep track of the distance between them [24]. In the cooperative localization, ranging technique is applied to measure the distance between multiple agents as a network to improve the overall navigation accuracy of each of the agent [39].

1.5 Outline of the Book

The topic of this book is about the pedestrian inertial navigation and related self-contained aiding techniques. In Chapter 2, we first introduce the technological basis of inertial navigation – inertial sensors, and IMUs. Their basic principles of operation, technology background, and state-of-the-art are included. Next, in Chapter 3, basic implementation and algorithm of strapdown inertial navigation are presented as a basis of the following analysis. Then, we demonstrate how the navigation errors are accumulated in the navigation process in Chapter 4, with a purpose of pointing out the importance of aiding in the pedestrian inertial navigation. Chapter 5 introduces one of the most commonly used aiding technique in pedestrian inertial navigation: ZUPT aiding algorithm. It is followed by an analysis on navigation error propagation in the ZUPT-aided pedestrian inertial navigation in Chapter 6, relating the navigation error to the IMU errors. Chapter 7 presents some of the limitations of the ZUPT-aided pedestrian inertial navigation, and methods have been proposed and demonstrated to be able to reduce the majority part of the errors caused by the ZUPTs. Chapter 8 discusses efforts in improving the adaptivity of the pedestrian inertial navigation algorithm. Approaches including ML and Multiple-Model (MM) methods are introduced. Chapter 9 discusses other popular self-contained aiding techniques, such as magnetometry, barometry, computer vision, and ranging techniques. Different ranging types, mechanisms, and implementations are covered in this chapter. Finally, in Chapter 10, the book concludes with a technological perspective on self-contained pedestrian inertial navigation with an outlook for development of the Ultimate Navigation Chip (uNavChip).

References

1 Bowditch, N. (2002). *The American Practical Navigator*, Bicentennial Edition. Bethesda, MD: National Imagery and Mapping Agency.
2 Sobel, D. (2005). *Longitude: The True Story of a Lone Genius Who Solved the Greatest Scientific Problem of His Time*. Macmillan.

3 Hofmann-Wellenhof, B., Lichtenegger, H., and Wasle, E. (2007). *GNSS-Global Navigation Satellite Systems: GPS, GLONASS, Galileo, and More*. Springer Science & Business Media.

4 Titterton, D. and Weston, J. (2004). *Strapdown Inertial Navigation Technology*, 2e, vol. 207. AIAA.

5 Woodman, O.J. (2007). An Introduction to Inertial Navigation. No. UCAM-CL-TR-696. University of Cambridge Computer Laboratory.

6 Wagner, J. and Trierenberg, A. (2010). The machine of Bohnenberger: bicentennial of the gyro with cardanic suspension. *Proceedings in Applied Mathematics and Mechanics* 10 (1): 659–660.

7 Prikhodko, I.P., Zotov, S.A., Trusov, A.A., and Shkel, A.M. (2012). Foucault pendulum on a chip: rate integrating silicon MEMS gyroscope. *Sensors and Actuators A: Physical* 177: 67–78.

8 Tazartes, D. (2014). An historical perspective on inertial navigation systems. *IEEE International Symposium on Inertial Sensors and Systems (ISISS)*, Laguna Beach, CA, USA (25–26 February 2014).

9 Ma, M., Song, Q., Li, Y., and Zhou, Z. (2017). A zero velocity intervals detection algorithm based on sensor fusion for indoor pedestrian navigation. *IEEE Information Technology, Networking, Electronic and Automation Control Conference (ITNEC)*, Chengdu, China (15–17 December 2017).

10 Perlmutter, M. and Robin, L. (2012). High-performance, low cost inertial MEMS: a market in motion!. *IEEE/ION Position, Location and Navigation Symposium*, Myrtle Beach, SC, USA (23–26 April 2012).

11 Jopling, P.F. and Stameris, W.A. (1970). Apollo guidance, navigation and control-design survey of the Apollo inertial subsystem.

12 Foxlin, E. (2005). Pedestrian tracking with shoe-mounted inertial sensors. *IEEE Computer Graphics and Applications* 25 (6): 38–46.

13 Harle, R. (2013). A survey of indoor inertial positioning systems for pedestrians. *IEEE Communications Surveys & Tutorials* 15 (3): 1281–1293.

14 Díez, L.E., Bahillo, A., Otegui, J., and Otim, T. (2018). Step length estimation methods based on inertial sensors: a review. *IEEE Sensors Journal* 18 (17): 6908–6926.

15 Köse, A., Cereatti, A., and Della Croce, U. (2012). Bilateral step length estimation using a single inertial measurement unit attached to the pelvis. *Journal of Neuroengineering and Rehabilitation* 9 (1): 1–10.

16 Miyazaki, S. (1997). Long-term unrestrained measurement of stride length and walking velocity utilizing a piezoelectric gyroscope. *IEEE Transactions on Biomedical Engineering* 44 (8): 753–759.

17 Bishop, E. and Li, Q. (2010). Walking speed estimation using shank-mounted accelerometers. *IEEE International Conference on Robotics and Automation*, Anchorage, AK, USA (3–7 May 2010).

18 Omr, M. (2015). Portable navigation utilizing sensor technologies in wearable and portable devices. PhD dissertation. Department of Electrical and Computer Engineering, Queens University.

19 Renaudin, V., Susi, M., and Lachapelle, G. (2012). Step length estimation using handheld inertial sensors. *Sensors* 12 (7): 8507–8525.

20 Munoz Diaz, E. (2015). Inertial pocket navigation system: unaided 3D positioning. *Sensors* 15 (4): 9156–9178.

21 Beauregard, S. (2006). A helmet-mounted pedestrian dead reckoning system. *VDE International Forum on Applied Wearable Computing*, Bremen, Germany (15–16 March 2006).

22 Park, J.-G., Patel, A., Curtis, D. et al. (2012). Online pose classification and walking speed estimation using handheld devices. *ACM Conference on Ubiquitous Computing*, New York City, NY, USA (September 2012).

23 Wang, Y., Jao, C.-S., and Shkel, A.M. (2021) Scenario-dependent ZUPT-aided pedestrian inertial navigation with sensor fusion. *Gyroscopy and Navigation* 12 (1).

24 Laverne, M., George, M., Lord, D. et al. (2011). Experimental validation of foot to foot range measurements in pedestrian tracking. *ION GNSS Conference*, Portland, OR, USA (19–23 September 2011).

25 Wang, Y., Lin, Y.-W., Askari, S. et al. (2020). Compensation of systematic errors in ZUPT-aided pedestrian inertial navigation. *IEEE/ION Position Location and Navigation Symposium (PLANS)*, Portland, OR, USA (20–23 April 2020).

26 TDK InvenSense (2020). ICP-10100 Barometric Pressure Sensor Datasheet.

27 Cui, X., Gulliver, T.A., Li, J., and Zhang, H. (2016). Vehicle positioning using 5G millimeter-wave systems. *IEEE Access* 4: 6964–6973.

28 Gersdorf, B. and Freese, U. (2013). A Kalman filter for odometry using a wheel mounted inertial sensor. *International Conference on Informatics in Control, Automation and Robotics (ICINCO)* (1), 388–395.

29 Jimenez, A.R., Seco, F., Prieto, J.C., and Guevara, J. (2010). Indoor pedestrian navigation using an INS/EKF framework for yaw drift reduction and a foot-mounted IMU. *IEEE Workshop on Positioning Navigation and Communication (WPNC)*, Dresden, Germany (11–12 March 2010).

30 Mezentsev, O. and Collin, J. (2019). Design and performance of wheel-mounted MEMS IMU for vehicular navigation. *IEEE International Symposium on Inertial Sensors & Systems*, Naples, FL, USA (1–5 April 2019).

31 Zheng, Y., Liu, Q., Chen, E. et al. (2014). Time series classification using multi-channels deep convolutional neural networks. *International Conference on Web-Age Information Management*, Macau, China (16–18 June 2014), pp. 298–310.

32 Hannink, J., Kautz, T., Pasluosta, C.F. et al. (2017). Mobile stride length estimation with deep convolutional neural networks. *IEEE Journal of Biomedical and Health Informatics* 22 (2): 354–362.

33 Wagstaff, B., Peretroukhin, V., and Kelly, J. (2017). Improving foot-mounted inertial navigation through real-time motion classification. *IEEE International Conference on Indoor Positioning and Indoor Navigation (IPIN)*, Sapporo, Japan (18–21 September 2017).

34 Fan, L., Wang, Z., and Wang, H. (2013). Human activity recognition model based on decision tree. *IEEE International Conference on Advanced Cloud and Big Data*, Nanjing, China (13–15 December 2013).

35 Wang, Y. and Shkel, A.M. (2021) Learning-based floor type identification in ZUPT-aided pedestrian inertial navigation. *IEEE Sensors Conference* 5.

36 Askari, S., Jao, C.-S., Wang, Y., and Shkel, A.M. (2019). Learning-based calibration decision system for bio-inertial motion application. *IEEE Sensors Conference*, Montreal, Canada (27–30 October 2019).

37 Anguita, D., Ghio, A., Oneto, L. et al. (2012). Human activity recognition on smartphones using a multiclass hardware-friendly support vector machine. In: *International Workshop on Ambient Assisted Living*, 216–223. Berlin, Heidelberg: Springer-Verlag.

38 Ordóñez, F.J. and Roggen, D. (2016). Deep convolutional and LSTM recurrent neural networks for multimodal wearable activity recognition. *Sensors* 16 (1): 115.

39 Olsson, F., Rantakokko, J., and Nygards, J. (2014). Cooperative localization using a foot-mounted inertial navigation system and ultrawideband ranging. *IEEE International Conference on Indoor Positioning and Indoor Navigation (IPIN)*, Busan, Korea (27–30 October 2014).

2

Inertial Sensors and Inertial Measurement Units

Inertial sensors (accelerometers and gyroscopes) are the hardware basis for inertial navigation. Inertial sensors are sensors based on inertia and relevant measuring principles. There are two types of inertial sensors: accelerometers and gyroscopes, measuring the specific forces and rotations, respectively. For navigation applications, the inertial sensors are typically assembled into Inertial Measurement Units (IMUs), which are comprised of three accelerometers and three gyroscopes mounted orthogonal to each other. In this chapter, we mainly focus on the inertial sensors and IMUs in context of their operating principles.

2.1   Accelerometers

Inertial navigation relies on the measurement of acceleration, which can be integrated into velocity and position change. According to Newton's second law of motion, the acceleration of a rigid body with respect to the inertial space is proportional to the force applied to it. Therefore, the measurement of acceleration can be conducted completely internally, and it does not require any external references, like in the measurement of velocity or position.

Accelerometers can typically be categorized into two classes: static accelerometers and resonant accelerometers. In static accelerometers, the sensing element, or the proof mass, is typically not vibrating during measurement, while the resonant accelerometers are actuated into their resonant frequencies during measurement [1].

2.1.1   Static Accelerometers

A static accelerometer can generally be modeled as a damped mass-spring system (Figure 2.1). The proof mass m is connected to the frame of the accelerometer by

Pedestrian Inertial Navigation with Self-Contained Aiding, First Edition. Yusheng Wang and Andrei M. Shkel.
© 2021 The Institute of Electrical and Electronics Engineers, Inc. Published 2021 by John Wiley & Sons, Inc.

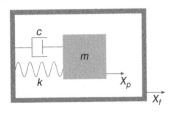

Figure 2.1 The basic structure of an accelerometer.

a spring with stiffness k and a damper with damping coefficient c. Suppose the displacement of the frame is x_f and the displacement of the proof mass relative to the absolute reference frame is x_p, then the equation of motion can be written as:

$$m\ddot{x}_p = c(\dot{x}_f - \dot{x}_p) + k(x_f - x_p). \tag{2.1}$$

Assume $x = x_p - x_f$ to be the relative motion of the proof mass with respect to the frame, which is the value to be measured, then the equation becomes:

$$m\ddot{x} + c\dot{x} + kx = -m\ddot{x}_f. \tag{2.2}$$

In this case, the relative motion of the proof mass is directly related to the acceleration of the frame. It can be proven that the relative motion of the proof mass is proportional to the external acceleration with a ratio of $1/\omega_0^2$, if the acceleration changes slowly:

$$\frac{x}{\ddot{x}_f} = \frac{m}{k} \triangleq \frac{1}{\omega_0^2}, \quad \text{if } \omega \ll \omega_0, \tag{2.3}$$

where ω is the frequency of the external acceleration.

Even though the modeling of accelerometers is as simple as a damped mass-spring system, the real implementation of accelerometers can be very different. For example, in the thermal accelerometers, only the ambient gas, a heater, and some spatially distributed thermometers are needed. The heater will heat up the gas and create a temperature distribution in the ambient gas, which is symmetric when the system is at rest. However, the temperature distribution will change due to the inertial forces applied to the heated gas when an external acceleration is applied, and the temperature change can be picked up by the thermometers around the heater. As a result, the ambient gas is used to replace the solid proof mass in the thermal accelerometers. Due to the absence of a solid suspended mass, the thermal accelerometers theoretically have a better robustness against shock, stiction, and environmental vibrations [2].

Fiber optic accelerometers have been demonstrated for acceleration measurement [3]. The operation mechanism is to take advantage of the change of the refractive index of the material due to the bend of the fiber caused by external accelerations. As a result, the phase of the output light will change. Fiber optic accelerometers have demonstrated good performance, especially for low frequency (<10 Hz) and weak vibration excitation (<0.3 m/s^2) measurement [4]. The sensor can be made immune to temperature fluctuation if using the interrogation scheme of power detection. Besides, like the thermal accelerometers, no moving parts are involved in the fiber optic accelerometers, thus the they are free from drifts due to shock and vibrations by nature.

Notice that in such systems, the gain of the accelerometer will increase with a lower natural frequency of the system, but the bandwidth will decrease correspondingly. In addition, the system input (the specific force) produces a proportional change in the sensor output voltage, i.e. the sensor input is Amplitude Modulated (AM). The dynamic range of such system is typically lower than 120 dB due to the fact that commercial available references for AM signals have a stability of about 1 ppm [5]. Besides, packaging of the device is designed such that the system is critically damped achieved by residual gas sealed in the package, in order to damp away the system's transient response. Such a packaging requirement contradicts the vacuum sealing requirements of high-performance micro-electro-mechanical sytems (MEMS) gyroscopes, and will complicate single die integration of the system [6]. Therefore, accelerometers based on Frequency Modulation (FM), or resonant accelerometers, have been proposed and demonstrated, where the frequency of the output, instead of the amplitude is detected and used as a measurement of the external acceleration. Higher dynamic range, bandwidth, and signal-to-noise ratio have been demonstrated.

2.1.2 Resonant Accelerometers

Unlike the static accelerometer whose output is directly related to the amplitude of the motion, the output of resonant accelerometer is the resonant frequency of the device, which is correlated to the external acceleration. A majority of the resonant accelerometers are designed in a way that the effective stiffness of the device will change due to the structural stress imposed by the external acceleration.

Different mechanisms can be utilized to achieve the goal. One of the earliest implementation is based on the Surface Acoustic Wave (SAW) devices [7], where a proof mass is fixed on the free end of a cantilever beam made of piezoelectric material, such as quartz or Lead Zirconate Titanate (PZT), so that the mechanical deformation can be transferred to electrical signals. When the external acceleration is applied, the cantilever beam will bend, the geometric dimension on the surface, as well as the elastic modulus and also the density of the material in the surface layer will change due to the stretching or compression. As a result, the resonant frequency of the SAW will change. A sensitivity on the order of 10 kHz/g can be achieved. One of the main advantages of SAW devices is that their fabrication is relatively easy and they have a relatively larger fabrication tolerance.

Vibrating beams can also be configured as resonant accelerometers. One end of the clamped-clamped beam can be connected to a proof mass through a lever. When an external acceleration is applied to the system, the inertial force created by the proof mass will be magnified by the lever, and applied on the beam as an axial load, which will change the resonant frequency of the beam. As a result, the magnitude of the acceleration can be extracted by measuring the resonant frequency of the beam. Performance as good as 56 ng of bias instability and 98 ng/$\sqrt{Hz}$ of Velocity Random Walk (VRW) has been demonstrated on micro-fabricated MEMS accelerometers [8].

A Bulk Acoustic Wave (BAW) device has also been demonstrated to operate as a resonant accelerometer in [9], where the resonant frequency of a BAW resonator can be electrostatically tuned. When an external acceleration is applied, the tether connecting the resonator to the anchor will deform due to the acceleration, and the capacitive gap between the resonator and the frequency tuning electrode will change. As a result, the frequency shift caused by the electrostatic tuning will change. In this case, however, the relation between the frequency shift and the external acceleration is not linear due to the nonlinear relation between the tuning capacitance and the gap. One mitigation is to change the design, such that instead of the gap, the overlapping area between the resonator and the tuning electrode change with external acceleration, and a linear dynamic range above 140 dB has been demonstrated [10]. One of the benefits of using BAW resonator is that it has a much higher resonant frequency than other devices, and therefore, a higher bandwidth and shock survivability can be achieved.

Schematics of accelerometers based on SAW devices, vibrating beams, and BAW devices are shown in Figure 2.2.

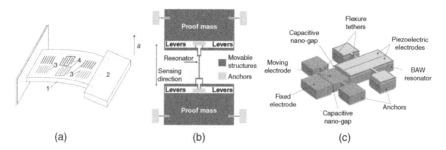

(a) (b) (c)

Figure 2.2 Schematics of accelerometers based on SAW devices [11], vibrating beams [8], and BAW devices [9]. Source: (a) Shevchenko et al. [11]. Licensed under CC BY 4.0, (b) Zhao et al. [8], (c) Daruwalla et al. [9].

2.2 Gyroscopes

Gyroscope is a kind of sensor that measures rotation. Main applications of gyroscopes include automotive rotation detection, platform stabilization, gyro-compassing, and inertial navigation. Gyroscopes can be categorized into different classes according to their operating physical principles and the involved technology. Some of the gyroscope classes include mechanical gyroscopes, optical gyroscopes, Nuclear Magnetic Resonance (NMR) gyroscopes, and MEMS vibratory gyroscopes. Their performances and applications are summarized in Figure 2.3.

2.2.1 Mechanical Gyroscopes

Mechanical gyroscope is also called spinning mass gyroscope. It is the oldest form of the gyroscope developed in the history, but it is still by far the most accurate gyroscope. The core element in a mechanical gyroscope is a spinning mass that rotates at high speed on a gimbaled frame. Other suspension techniques, such as electrostatic and magnetic, have also been demonstrated. Mechanical gyroscopes operate according to the gyroscopic effect, where the axis of the rotation remains fixed with respect to the inertial frame in the absence of external torques due to conservation of angular momentum. Therefore, mechanical gyroscopes can directly measure the change of orientation of the system by measuring the relative orientation between the spinning mass and the frame. Mechanical gyroscopes typically have bias instabilities on the order of 10^{-5} to 10^{-3} °/h. Main error sources in the

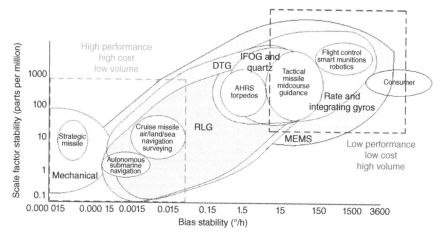

Figure 2.3 Typical performances and applications of different gyroscopes. Passaro et al. [12]. Licensed under CC BY 4.0.

mechanical gyroscopes are the friction and unbalance of the system. Mitigation strategies include high-precision fabrication and post-processing process, better bearings and lubricants, and even fluid or magnetic suspensions in some special cases. Despite the ultrahigh accuracy of the mechanical gyroscopes, their complicated system, large size and weight, and high cost limit the application mainly to submarine navigation.

2.2.2 Optical Gyroscopes

The basic operating principle of optical gyroscopes is the Sagnac effect, where a difference in the propagation time of two counter-propagating beams is induced by the rotation of the optical path. A corresponding phase shift between the two beams can be written as:

$$\Delta\phi = \frac{8\pi A}{\lambda c}\Omega, \tag{2.4}$$

where A is the area enclosed by the light path, λ is the wavelength of the beam, c is the speed of light, and Ω is the external angular rate. The Sagnac effect can be utilized in gyroscopes since the output – the phase difference, or equivalently the frequency difference – is proportional to the external angular rate according to (2.4). Besides, (2.4) indicates that a larger enclosed area and the shorter wavelength of the light will increase the sensitivity of the optical gyroscopes. There are currently two types of mature optical gyroscopes that are commercially available: Ring Laser Gyroscopes (RLGs) and Fiber Optic Gyroscopes (FOGs).

2.2.2.1 Ring Laser Gyroscopes
RLGs typically consist of a solid block made of ceramic glass material as the lasing cavity, into which a helium/neon mixture is introduced as the lasing medium. The electrodes in the system provide gain for the lasing medium, and two independent beams will be sustained around the cavity of opposite directions. The output of the RLGs is obtained by interfering the two beams to create fringe patterns, which can be measured by a photo detector. The geometry of the lasing cavity should be accurate enough (submicron level) to make sure that the optical path in the cavity is a multiple of the wavelength, so that a standing wave can be generated. Bias instability of on the order of 0.0001 °/h has been demonstrated on RLGs [13]. Mechanical gyroscopes and the RLGs are the only two types of gyroscopes that are available on the high-end gyroscope market.

Lock-in effect is one of the main error sources in the RLGs, where the two counter-propagating laser beams are coupled and synchronized if the input rotation rate is low. The lock-in effect is mainly caused by the backscattering of the lasers due to the imperfections of lasing cavity. A common way to overcome it is to use mechanical dither to create a mechanical oscillation around the sense axis [14]. However, the dithering vibrations will introduce extra disturbance to the optical cavity, and thus increase the noise of the overall system.

The cost of RLGs is high due to the stringent requirement on the fabrication of lasing cavity. It is difficult to reduce the size and weight while keep the high performances. The power consumption of RLGs is high (typically around 10 W) to support the laser. FOGs are able to overcome these disadvantages at the cost of lower performances.

2.2.2.2 Fiber Optic Gyroscopes

Instead of solid bulk lasing cavities, FOGs take advantage of optical fiber technology to form the optical path, where two independent beams propagate in the opposite directions. The advantages of using optical fibers include the following: (i) the optical fibers can be formed into coils such that a long optical path can be achieved within the small sizes of the devices; (ii) the mass production of fiber components primarily for the telecommunication industry makes the cost of optical fibers affordable; (iii) the requirement on the fabrication tolerance of optical fibers in FOGs is much less stringent than that of the lasing cavities in the RLGs; and (iv) no mechanical dither is necessary to overcome the lock-in effect in the FOGs, making the system less complicated. More details on the comparison between the RLGs and FOGs are available in [15].

There are mainly two types of FOGs: interferometric FOG (I-FOG) and resonant FOG (R-FOG). I-FOG is the most mature type, where the phase difference between the counter-propagation beams due to the Sagnac effect is detected by measuring their interference, whereas R-FOG directly measure the frequency difference between the two beams.

In the I-FOGs, low-coherence light sources, such as superluminescent diode, are most commonly used to reduce the backscattering. As a result, the power consumption of I-FOGs is much lower than RLGs with a value on the order of 1 mW [16]. The length of the optical fibers can be on the order of kilometers to increase the gyroscope sensitivity. However, it suffers from bias drift due to the time variant temperature distribution, light source intensity noises, and polarization coupling noise. The fiber optic configuration also suffers more from shock and vibrations than the bulk optical cavities. A bias instability as low as 0.01°/h has been demonstrated [17].

With lower performances compared to the RLGs, the FOGs have found application in robotics and automotive industries, whereas the RLGs have been widely used in strapdown navigation systems in commercial and military aircraft.

2.2.3 Nuclear Magnetic Resonance Gyroscopes

Large-scale Nuclear Magnetic Resonance Gyroscopes (NMRGs) were first developed in the 1960s with bias instability of $0.1°/h$ [18], but its size was too large to be practically used. However, micro-scaled NMRGs have become possible with the development of MEMS-based micro and batch fabrications [19]. NMRGs take advantage of the Larmor precession, where the axes of the nuclear spin of atoms precess along the direction of the external magnetic field, and the external rotation will change the precessing frequency of the atoms. The observed precession frequency can be expressed as:

$$\omega_{\text{obs}} = \gamma H + \Omega, \tag{2.5}$$

where γ is the ratio of the magnetic dipole moment to the angular momentum of the nuclei, and it being dependent solely on the atom property, H is the magnitude of the external magnetic field, and Ω is the angular rate of the external rotation.

In a typical NMRG implementation, a vapor cell is used to contain NMR active isotopes, alkali atoms, and buffer gas. The nuclear spin polarization of NMR isotopes is hard to control and measure, but the angular momentum can be transferred between the nuclear spin of NMR isotopes and the electron spin of the alkali atoms by the collisions between the atoms. The buffer gas is used to reduce the collisions between the alkali atoms and the wall of the cell, thus minimize the Coherent Population Trapping (CPT) signal broadening induced by it [20]. Notice that the output angular rate is directly related to the magnitude of the magnetic field. As a result, the stability of the magnetic field is one of the most critical parameters that affect the performance of an NMRG. For example, the magnetic field instability of $100\,\text{fT}$, which is approximately one billionth the magnitude of the Earth's magnetic field, will introduce gyroscope bias instability of about $1°/h$, for ^{129}Xe [18]. Therefore, magnetic shielding and an accurate control of the imposed magnetic field are critical to improve the performance of NMRGs.

Theoretically, NMRGs have zero temperature sensitivity and unlimited linear dynamic range due to their atomic nature. Besides, their performance is not related to the size as in the case of optical gyroscopes. In addition, there is no moving part in NMRGs, making them inherently resistant to shocks and vibrations. The many advantages of NMRGs make this technology promising, especially after combined with the MEMS technology to achieve the miniaturization.

2.2.4 MEMS Vibratory Gyroscopes

The first micromachined silicon gyroscope was developed in the Charles Stark Draper Laboratory in 1991 [21]. Due to the rapid development of MEMS technology in the past three decades, MEMS gyroscopes are becoming smaller, more accurate, and lower in cost. MEMS gyroscope is the most used type of gyroscopes in the pedestrian inertial navigation. More details on the MEMS vibratory gyroscopes can be found in [22, 23].

2.2.4.1 Principle of Operation

Most MEMS gyroscopes use a vibrating mechanical element as the sensing element to detect the angular velocity. Compared to mechanical gyroscopes, MEMS gyroscopes do not have the spinning rotor or the complicated gimbal system, and therefore, small size and weight can be achieved. The operating principle in MEMS vibratory gyroscopes is the Coriolis effect: the object in a rotating frame will experience Coriolis force whose magnitude is proportional to the angular rate of the frame:

$$F_c = -2m\Omega \times v, \tag{2.6}$$

where m is the mass of the object, Ω is the angular rate of the frame, and v is the relative velocity of the object with respect to the frame. Notice that the magnitude of the Coriolis force, or equivalent the Coriolis acceleration, is proportional to the angular rate. As a result, the MEMS vibratory gyroscopes can be considered as a resonator combined with an accelerometer, where the resonator oscillates at resonance with a constant amplitude, and the accelerometer measures the magnitude of the Coriolis acceleration.

Different materials, such as Single Crystal Silicon (SCS), poly-silicon, metallic glass, Fused Quartz (FQ), diamond, and Silicon Carbide (SiC), have been explored to fabricate MEMS gyroscopes. Different geometries and configurations have also been demonstrated, including discrete mass-spring system, continuous ring structure, tuning fork, BAW device, and 3D shell structure. Although there are many different configurations of MEMS vibratory gyroscopes, they generally can be modeled as a Two Degree-of-Freedom (2-DOF) damped mass-spring system, as shown in Figure 2.4. In most MEMS gyroscopes, the resonant frequencies of the two modes are designed to be different due to technological limitations.

2.2.4.2 Mode of Operation

There are three operational modes in which a MEMS gyroscope can be controlled: open-loop mode, force-to-rebalance mode, and whole angle mode [28], and their ideal responses are shown in Figure 2.5.

Open-Loop Mode Open-loop mode is the simplest gyroscope operational mode, where the proof mass is actuated at resonance with a constant amplitude along the drive axis (x-axis), and no actuation or control is applied along the sense axis (y-axis). If the gyroscope experiences angular rate along its sensitive axis (z-axis in this case), Coriolis force will take effect and drive the y-axis. It can be shown

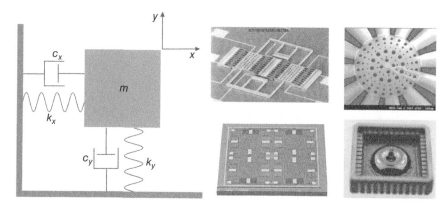

Figure 2.4 Schematics of a gyroscope and its different configurations [24]–[27]. Source: (a) Nasiri [24], (b) Johari and Ayazi [26], (d) Asadian et al. [27].

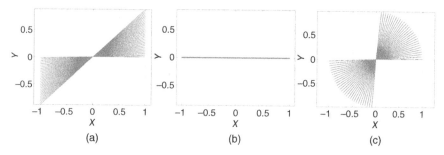

Figure 2.5 Ideal response of a gyroscope operated in (a) open-loop mode, (b) force-to-rebalance mode, and (c) whole angle mode, respectively. Source: Based on Shkel [28].

that the amplitude of motion along the y-axis is proportional to the angular rate along the z-axis. Open-loop implementation is simple at the cost of small gyroscope bandwidth and low linear measuring range.

Force-to-Rebalance Mode Force-to-rebalance mode has been developed to overcome the limits of open-loop mode. In force-to-rebalance mode, the control along the drive axis is the same as in the open-loop mode, while the amplitude of motion along the sense axis is suppressed to zero by applying the appropriate drive force. The controlling force to suppress the motion is recorded, and it is proportional to the input angular rate. The bandwidth of the gyroscope can be increased to the resonant frequency of the device, at the cost of more complicated control and reduced signal-to-noise ratio.

Whole Angle Mode In the whole angle operational mode, the proof mass vibrates freely in the x–y plane, and the orientation of the oscillation will precess due to the

Coriolis effect if there is input rotation along the z-axis. Unlike the previous two operational modes, the output of the gyroscopes operated in the whole angle mode is angle, instead of angular rate. As a result, the need for numerical integration of the angular rate into orientation is eliminated during navigation. Besides, gyroscopes operated in the whole angle mode have theoretically unlimited mechanical bandwidth, allowing them to work under severe dynamics. However, in order to be operated in the whole angle mode, the two resonant frequencies of the gyroscope need to be close to each other, which imposes stringent requirement on the fabrication and postprocessing of the devices. Therefore, despite many of the unique advantages of whole angle operational mode, it is still only reserved for precision machined macro-scale gyroscopes, and not available for MEMS vibratory gyroscopes. Some perspective of development of such gyroscopes can be found in [23].

2.2.4.3 Error Analysis

Major error sources in MEMS vibratory gyroscopes include drifts due to the change of the ambient temperature, electronic noises, and the cross-talk between similar sensors mounted on the same substrate, for example in an IMU.

Most MEMS vibratory gyroscopes are made of silicon, whose material properties, such as elastic modulus and coefficient of thermal expansion, change with respect to the temperature. Besides, the gains of some lost-cost operational amplifiers also change with temperature. Scale factor drift as high as $12\,000$ ppm/°C, and bias drift of 180 (°/h)/°C have been demonstrated on a silicon gyroscope [29]. In addition, the requirement on the small size and low power consumption also prevents the MEMS devices from using ovenization to control the ambient temperature. Therefore, the changing temperature is one of the dominant factors that affect the performance of MEMS gyroscopes. Mitigation methods include temperature self-sensing and self-calibration, where the temperature of the device is monitored by keeping track of its resonant frequency, and the output can be compensated according to the prerecorded calibration results. A temperature self-sensing with accuracy of 0.0004 °C, total bias of 2°/h, and scale factor stability of 700 ppm were achieved over temperature range of 25–55 °C [30]. Building the gyroscopes with materials of lower temperature-dependent parameters, such as FQ, is another possible method to overcome this issue.

Many control loops are involved in the MEMS vibratory gyroscopes to tune the frequencies of the device, sustain the oscillation at resonance, and suppress quadrature component of the motion. The noise level in the electronics may be high due to the restriction of size and power consumption of the overall system. Better electronics design may help to mitigate the problem, as demonstrated in [31]. Another approach is to reduce the as-fabricated imperfections of the devices, such that less amount of control is necessary. Various methods to compensate and reduce the structural asymmetry and energy dissipation of the devices have been developed, in order to improve the noise performance of the MEMS vibratory gyroscopes.

To summarize, due to the development of micro-fabrication technologies, MEMS vibratory gyroscopes have demonstrated much smaller size (footprint on the order of millimeter) than the aforementioned types of gyroscopes, while showing a moderate performance. Development of more accurate fabrication processes, implementation of better structural designs, new materials with lower energy dissipation, postprocessing techniques, and control electronics with lower noises boost the performance of MEMS vibratory gyroscopes. Near navigation grade performance with bias instability of 0.027 °/h and Angle Random Walk (ARW) of 0.0062 °/$\sqrt{h}$ have been demonstrated [32]. Currently, MEMS vibratory gyroscopes are mostly used in consumer and low-end tactical grade applications, but they are envisioned to be widely used in some of the high-end application in the near future.

2.3 Inertial Measurement Units

A single accelerometer or a single gyroscope is not enough for navigation applications. Typically, an IMU, which consists of three accelerometers and three gyroscopes, is needed to obtain the full information for navigation. In this section, we briefly introduce some of the commonly used technologies to combine individual inertial sensors into a single IMU.

2.3.1 Multi-sensor Assembly Approach

The most commonly used and mature technology is based on assembly of single inertial sensors on a rigid frame. Off-the-shelf single-axis inertial sensors are assembled along the three orthogonal directions to measure the acceleration and angular rate in the 3D space. One of the main advantages of this approach is that the individual inertial sensors can be optimized to improve the sensor performance and reject the off-axis input, due to less constraints as compared to the single-chip approaches [33]. For example, the out-of-plane motion that is necessary in the x- and y-axis gyroscopes and z-axis accelerometers can be eliminated in the cubic IMUs shown in Figure 2.6a. An alternative approach of stacking the structures is also popular in order to reduce the overall size of the IMU (Figure 2.6b).

Another advantage of the multi-sensor assembly approach is the reduction of cross-talk. Cross-talk in MEMS gyroscopes is the result of mechanical and electrical coupling between devices placed close to each other. Mechanical coupling is mainly due to the vibration transmission between the devices through the

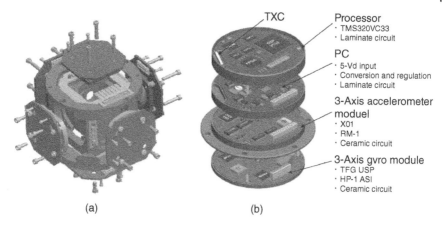

(a)   (b)

Figure 2.6 Schematics of two typical IMU assembly architectures: (a) cubic structure and (b) stacking structure. Source: (a) Based on Barbour et al. [34].

mounting structures, and the electrical coupling originates from the capacitive or resistive coupling through the close-spaced wire bonds and metal traces [35]. Since the inertial sensors are individually assembled on a rigid frame, the coupling between the sensor can be minimized if designed properly.

The main drawback of the multi-sensor assembly approach is the relatively large volume and weight of the system, since it requires an extra frame to fix the sensors, and an assembly of six separate Printed Circuit Boards (PCB) to process the readouts from the six inertial sensors.

Typically, IMUs requiring mid-to-high performance are fabricated using the multi-sensor assembly approach at the cost of relatively larger volumes. Some examples include Honeywell HG1930 [36], Systron Donner SDI500 [37], and Northrop Grumman μIMU [38]. Their volumes are typically on the order of 100 cm³ to achieve better performances. A miniaturized version of the multi-sensor assembly approach has been explored, experimentally demonstrating an IMU with volume of 10 mm³ (without control circuits) [39]. However, the performance of such IMU is not satisfactory for high-end applications.

2.3.2 Single-Chip Approach

In the single-chip approach, multiple multi-axis or single-axis sensors are fabricated on a single chip. The largest advantage of this approach is its ability to achieve small volume due to the compact design. Volume of less than 10 mm³ is typical for IMUs fabricated in the single-chip approach. However, the extremely small volume comes at the cost of suboptimal performance. For example, in multi-axis sensors, a single proof mass may be involved in different

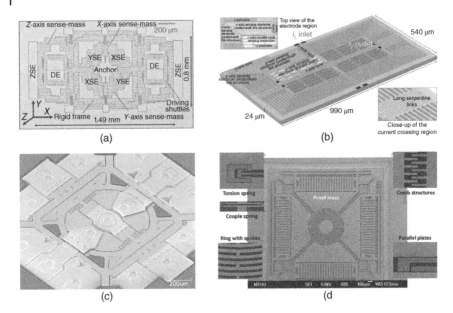

Figure 2.7 Different mechanical structures of three-axis gyroscopes. Source:
(a) Efimovskaya et al. [40], (b) Marra et al. [41–43].

translational and rotational motions in order to achieve measurements along
different directions, where the cross-talk between different axes may affect
the overall performance of the device. Figure 2.7 shows different mechanical
structures of three-axis gyroscopes. The footprints of such devices are generally
small (on the order of 1 mm²). However, the structures are typically complicated
to achieve the measurements along multiple directions, and as a result, the
reliability and performance may not be as good as their assembled counterparts.
Some commercially available IMUs fabricated in the single-chip approach include
Analog Devices ADIS16495 [44], STMicroelectronics LSM6DSM [45], and TDK
InvenSense ICM-42605 [46].

2.3.3 Device Folding Approach

The micro-fabricated version of the IMUs of cubic structure can be built in the
device folding approach, where single-axis high-performance inertial sensors can
be batched-fabricated on a flat wafer, and then the flat structure is folded into
a 3D configuration, like a silicon-origami, by the connecting flexible polymer
hinges [33]. In this approach, the whole IMU structure can be fabricated from
a single wafer, thus simplifying the fabrication process, and the folding process
can be assisted by jigs to improve both the yield and the accuracy. For example,

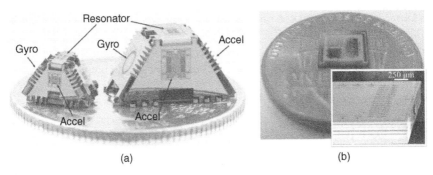

Figure 2.8 Examples of miniaturized IMU assembly architectures by MEMS fabrication: (a) folded structure and (b) stacking structure. Source: (a) Efimovskaya et al. [47], (b) Cao et al. [48].

Figure 2.8a shows an IMU prototype with three gyroscopes, three accelerometers, and one resonator as a clock. The main concerns in this approach include the compatibility of different materials (silicon, polymer, and metals) during fabrication, survivability of the structure under shock and vibration, inertial sensor misalignment during assembly, and temperature sensitivity of the structure. Structural reinforcement techniques, such as silicon welding and eutectic bonding between the latches, and new materials, such as Parylene, can be used to mitigate some of the concerns [49]. No commercial IMUs are currently available using this approach.

2.3.4 Chip-Stacking Approach

Another IMU miniaturization approach is through vertical chip stacking. IMUs fabricated in this approach have a similar form as shown in Figure 2.6b, except that each layer is fabricated out of a single chip and they are stacked together through micro-fabrication bonding instead of mechanical screws. This approach takes advantage of the fact that the inertial sensors are mostly planar whose thickness is much smaller than the length and width. Therefore, the chip-stacking approach will greatly reduce the overall volume of the entire system [50]. There are two main technologies that are necessary for the chip-stacking approach: wafer bonding technology and through-wafer interconnect technology. The purpose of wafer bonding is to form the rigid structure of the chip-stack, where the bonding yield and the misalignment are typically of great concern. The through-wafer interconnects are mainly for electrical connection between the stacks, and the main concerns are their aspect ratio, conductivity, and the feed-through between different interconnects [51]. IMUs fabricated in the chip-stacking approach have been demonstrated to have a volume as small as

13 mm^2 with a foot print of 6 mm × 6 mm (see Figure 2.8b) [48]. Compared to the IMUs fabricated in the single-chip approach, the volume is on the same order, while the large footprint indicates the potential of further improving the sensor performance. No commercial IMUs have been fabricated to date in this process mainly due to the relatively low reliability of the fabrication process.

2.4 Conclusions

In this chapter, we introduced some basic operating principles of different types of inertial sensors and IMUs. Different technologies are involved in the fabrication of individual inertial sensors and IMUs, and they all have their advantages and disadvantages.

Even though the overall topic of this book is about inertial navigation, where IMUs are used for navigation purposes, many other applications of IMUs exist, such as manufacturing quality control, medical rehabilitation, robotics, sports learning, and virtual and augmented reality [52]. Different constraints exist for different applications, but some common considerations include the package size, data accuracy, sampling rate, and environmental resistance. There is no IMU technology that is best for all applications, and therefore, a proper selection of the technology is needed for various application scenarios.

References

1 Ayazi, F. (2013). Bulk acoustic wave accelerometers. US Patent No. 8,528,404, 10 September 2013.

2 Everhart, C.L.M., Kaplan, K.E., Winterkorn, M.M. et al. (2018). High stability thermal accelerometer based on ultrathin platinum ALD nanostructures. *IEEE International Conference on Micro Electro Mechanical Systems (MEMS)*, Belfast, UK (21–25 January 2018), pp. 976–979.

3 Villnow, M. (2018). Fiber-optic accelerometer. US Patent Application 15/758,422, 13 September 2018.

4 Rong, Q., Guo, T., Bao, W. et al. (2017). Highly sensitive fiber-optic accelerometer by grating inscription in specific core dip fiber. *Scientific Reports* 7 (1): 11856.

5 Analog Devices (2008). MT-087: voltage references. Norwood, MA, USA. https://www.analog.com/media/en/training-seminars/tutorials/MT-087.pdf (accessed 05 March 2021).

6 Trusov, A.A., Zotov, S.A., Simon, B.R., and Shkel, A.M. (2013). Silicon accelerometer with differential frequency modulation and continuous

self-calibration. *IEEE International Conference on Micro Electro Mechanical Systems (MEMS)*, Taipei, Taiwan (20–24 January 2013).

7 Dwyer, D.F.G. and Bower, D.E. (1986). Surface acoustic wave accelerometer. US Patent No. 4,598,587, 8 July 1986.

8 Zhao, C., Pandit, M., Sobreviela, G. et al. (2019). A resonant MEMS accelerometer with 56ng bias stability and 98ng/Hz1/2 noise floor. *IEEE/ASME Journal of Microelectromechanical Systems* 28 (3): 324–326.

9 Daruwalla, A., Wen, H., Liu, C.-S. et al. (2018). A piezo-capacitive BAW accelerometer with extended dynamic range using a gap-changing moving electrode. *IEEE/ION Position, Location and Navigation Symposium (PLANS)*, Monterey, CA, USA (23-26 April 2018), pp. 283–287.

10 Shin, S., Daruwalla, A., Gong, M. et al. (2019). A piezoelectric resonant accelerometer for above 140db linear dynamic range high-G applications. *20th International Conference on Solid-State Sensors, Actuators and Microsystems & Eurosensors XXXIII (TRANSDUCERS & EUROSENSORS XXXIII)*, Berlin, Germany (23–27 June 2019), pp. 503–506.

11 Shevchenko, S., Kukaev, A., Khivrich, M., and Lukyanov, D. (2018). Surface-acoustic-wave sensor design for acceleration measurement. *Sensors* 18 (7): 2301.

12 Passaro, V., Cuccovillo, A., Vaiani, L. et al. (2017). Gyroscope technology and applications: a review in the industrial perspective. *Sensors* 17 (10): 2284.

13 Schreiber, K.U. and Wells, J.P.R. (2013). Invited review article: large ring lasers for rotation sensing. *Review of Scientific Instruments* 84 (4): 041101.

14 Fan, Z., Luo, H., Lu, G., and Hu, S. (2012). Direct dither control without external feedback for ring laser gyro. *Optics & Laser Technology* 44 (4): 767–770.

15 Juang, J. and Radharamanan, R. (2009). Evaluation of ring laser and fiber optic gyroscope technology. *American Society for Engineering Education, Middle Atlantic Section ASEE Mid-Atlantic Fall 2009 Conference*, King of Prussia, PA, USA (23-24 October 2009).

16 Nayak, J. (2011). Fiber-optic gyroscopes: from design to production. *Applied Optics* 50 (25): E152–E161.

17 Yu, Q., Li, X., and Zhou, G. (2009). A kind of hybrid optical structure IFOG. *International Conference on Mechatronics and Automation*, Changchun, China (9-12 August 2009), pp. 5030–5034.

18 Donley, E.A. (2010). Nuclear magnetic resonance gyroscopes. *IEEE Sensors Conference*, Kona, HI, USA (1-4 November 2010).

19 Larsen, M. and Bulatowicz, M. (2012). Nuclear magnetic resonance gyroscope: for DARPA's micro-technology for positioning, navigation and timing program. *IEEE International Frequency Control Symposium Proceedings*, Baltimore, MD, USA (21-24 May 2012), pp. 1–5.

20 Hasegawa, M., Chutani, R.K., Gorecki, C. et al. (2011). Microfabrication of cesium vapor cells with buffer gas for MEMS atomic clocks. *Sensors and Actuators A: Physical* 167 (2): 594–601.

21 Greiff, P., Boxenhorn, B., King, T., and Niles, L. (1991). Silicon monolithic micromechanical gyroscope. *IEEE International Conference on Solid-State Sensors and Actuators (TRANSDUCERS)*, San Francisco, CA, USA (24-27 June 1991), pp. 966–968.

22 Acar, C. and Shkel, A.M. (2008). *MEMS Vibratory Gyroscopes: Structural Approaches to Improve Robustness*. Springer Science & Business Media.

23 Senkal, D. and Shkel, A.M. (2020). *Whole Angle MEMS Gyroscopes: Challenges and Opportunities*. Wiley.

24 Nasiri, S. (2005). *A Critical Review of MEMS Gyroscopes Technology and Commercialization Status*. InvenSense.

25 Trusov, A.A., Atikyan, G., Rozelle, D.M. et al. (2014). Flat is not dead: current and future performance of Si-MEMS quad mass gyro (QMG) system. *IEEE/ION Position, Location and Navigation Symposium (PLANS)*, Monterey, CA, USA (5-8 May 2014), pp. 252–258.

26 Johari, H. and Ayazi, F. (2007). High-frequency capacitive disk gyroscopes in (100) and (111) silicon. *IEEE 20th International Conference on Micro Electro Mechanical Systems (MEMS)*, Hyogo, Japan (21-25 January 2007), pp. 47–50.

27 Asadian, M.H., Wang, Y., and Shkel, A.M. (2019). Development of 3D fused quartz hemi-toroidal shells for high-Q resonators and gyroscopes. *IEEE Journal of Microelectromechanical Systems* 28 (6): 954–964.

28 Shkel, A.M. (2006). Type I and type II micromachined vibratory gyroscopes. *IEEE/ION Position, Location, And Navigation Symposium (PLANS)*, Coronado, CA, USA (25-27 April 2006), pp. 586–593.

29 Prikhodko, I.P., Zotov, S.A., Trusov, A.A., and Shkel, A.M. (2012). Thermal calibration of silicon MEMS gyroscopes. *IMAPS International Conference and Exhibition on Device Packaging*, Scottsdale, AZ, USA (5-8 March 2012).

30 Prikhodko, I.P., Trusov, A.A., and Shkel, A.M. (2013). Compensation of drifts in high-Q MEMS gyroscopes using temperature self-sensing. *Sensors and Actuators A: Physical* 201: 517–524.

31 Wang, D., Efimovskaya, A., and Shkel, A.M. (2019). Amplitude amplified dual-mass gyroscope: design architecture and noise mitigation strategies. *IEEE International Symposium on Inertial Sensors and Systems*, Naples, FL, USA (1–5 April 2019).

32 Singh, S., Woo, J.-K., He, G. et al. (2020). $0.0062\,°/\sqrt{hr}$ Angle random walk and $0.027\,°/hr$ bias instability from a micro-shell resonator gyroscope with surface electrodes. *IEEE 33rd International Conference on Micro Electro Mechanical Systems (MEMS)*, Vancouver, Canada (18-22 January 2020), pp. 737–740.

33 Efimovskaya, A., Lin, Y.-W., and Shkel, A.M. (2017). Origami-like 3-D folded MEMS approach for miniature inertial measurement unit. *IEEE Journal of Microelectromechanical Systems* 26 (5): 1030–1039.

34 Barbour, N., Hopkins, R., Connelly, J. et al. (2010). Inertial MEMS system applications. NATO RTO Lecture Series. *RTO-EN-SET-116*. Low-Cost Navigation Sensors and Integration Technology.

35 Efimovskaya, A., Lin, Y.-W., Yang, Y. et al. (2017). On cross-talk between gyroscopes integrated on a folded MEMS IMU Cube. *IEEE 30th International Conference on Micro Electro Mechanical Systems (MEMS)*, Las Vegas, NV, USA (22-26 January 2017), pp. 1142–1145.

36 Honyywell (2018). HG1930 Inertial Measurement Unit. https://aerospace .honeywell.com/en/learn/products/sensors/hg1930-inertial-measurement-unit (accessed 05 March 2021).

37 EMCORE (2020). SDI500 Tactical Grade IMU Inertial Measurement Unit. https://emcore.com/products/sdi500-tactical-grade-imu-inertial-measurement-unit/ (accessed 08 March 2021).

38 Northrop Grumman LITEF GmbH (2016). μIMU Micro Inertial Measurement Unit. https://northropgrumman.litef.com/en/products-services/industrial-applications/product-overview/mems-imu/ (accessed 08 March 2021).

39 Zhu, W., Wallace, C.S., and Yazdi, N. (2016). A tri-fold inertial measurement unit fabricated with a batch 3-D assembly process. *IEEE International Symposium on Inertial Sensors and Systems*, Laguna Beach, USA (22-25 February 2016).

40 Efimovskaya, A., Yang, Y., Ng, E. et al. (2017). Compact roll-pitch-yaw gyroscope implemented in wafer-level epitaxial silicon encapsulation process. *IEEE International Symposium on Inertial Sensors and Systems (INERTIAL)*, Kauai, HI, USA (27-30 March 2017).

41 Marra, C.R., Gadola, M., Laghi, G. et al. (2018). Monolithic 3-axis MEMS multi-loop magnetometer: a performance analysis. *IEEE Journal of Microelectromechanical Systems* 27 (4): 748–758.

42 Wen, H., Daruwalla, A., Liu, C.-S., and Ayazi, F. (2018). A high-frequency resonant framed-annulus pitch or roll gyroscope for robust high-performance single-chip inertial measurement units. *IEEE Journal of Microelectromechanical Systems* 27 (6): 995–1008.

43 Tseng, K.-J., Li, M.-H., and Li, S.-S. (2020). A monolithic tri-axis MEMS gyroscope operating in air. *IEEE International Symposium on Inertial Sensors and Systems (INERTIAL)*, Hiroshima, Japan (23-26 March 2020).

44 Analog Devices (2020). ADIS16495 Tactical Grade, Six Degrees of Freedom IMU. https://www.analog.com/en/products/adis16495.html (accessed 08 March 2021).

45 STMicroelectronics (2017). iNEMO 6DoF Inertial Measurement Unit. https://www.st.com/en/mems-and-sensors/lsm6dsm.html (accessed 08 March 2021).

46 TDK InvenSense (2020). High Performance Low Power 6-Axis MEMS Motion Sensor. https://invensense.tdk.com/products/motion-tracking/6-axis/icm-42605/ (accessed 08 March 2021).

47 Efimovskaya, A., Senkal, D., and Shkel, A.M. (2015). Miniature origami-like folded MEMS TIMU. *IEEE International Conference on Solid-State Sensors, Actuators and Microsystems (TRANSDUCERS)*, Anchorage, AK, USA (21-25 June 2015).

48 Cao, Z., Yuan, Y., He, G. et al. (2013). Fabrication of multi-layer vertically stacked fused silica microsystems. *Transducers & Eurosensors XXVII: The 17th International Conference on Solid-State Sensors, Actuators and Microsystems (TRANSDUCERS & EUROSENSORS XXVII)*, Barcelona, Spain (16-20 June 2013).

49 Lin, Y.-W., Efimovskaya, A., and Shkel, A.M. (2017). Study of environmental survivability and stability of folded MEMS IMU. *IEEE International Symposium on Inertial Sensors and Systems (INERTIAL)*, Kauai, HI, USA (27-30 March 2017).

50 Duan, X., Cao, H., and Liu, Z. (2017). 3D stack method for micro-PNT based on TSV technology. *IEEE 3rd Information Technology and Mechatronics Engineering Conference (ITOEC)*, Chongqing, China (3-5 October 2017).

51 Efimovskaya, A., Lin, Y.-W., and Shkel, A.M. (2018). Double-sided process for MEMS SOI sensors with deep vertical Thru-Wafer interconnects. *IEEE Journal of Microelectromechanical Systems* 27 (2): 239–249.

52 Ahmad, N., Ghazilla, R.A.R., Khairi, N.M., and Kasi, V. (2013). Reviews on various inertial measurement unit (IMU) sensor applications. *International Journal of Signal Processing Systems* 1 (2): 256–262.

3

Strapdown Inertial Navigation Mechanism

Strapdown inertial navigation systems are now the most common form of inertial navigation system (INS) due to its potential benefits of lower cost, reduced size, and greater reliability compared with equivalent gimbal systems [1]. Due to these advantages, all current pedestrian inertial navigation systems take the architecture of strapdown inertial navigation. This chapter introduces fundamentals on the strapdown inertial navigation mechanism. Further material can be found in [1–3].

3.1 Reference Frame

Reference frames are needed to describe the motion of an object. It is of fundamental importance in inertial navigation to have a proper and precise definition of reference frames, since inertial navigation is independent of any external environments. The commonly used Cartesian coordinate frames include the following:

- Inertial frame (i-frame) is a nonrotating frame with respect to the fixed stars. Its origin is at the center of the Earth, and z-axis coincides with the Earth's polar axis. Although the Earth is moving with respect to the Sun, for measurements made in the vicinity of the Earth, i-frame can be considered to be an ideal inertial frame where Newton's laws of motion can be directly applied.
- Earth frame (e-frame) is a rotating frame. Its origin is at the center of the Earth and axes fixed with respect to the Earth. Typically, its one axis coincides with the Earth's polar axis and one axis is in the plane defined by the Greenwich meridian. Strictly speaking, e-frame is a non-inertial frame due to the Earth's rotation, but it is usually more convenient to use e-frame than i-frame for navigation on or around the Earth.
- Navigation frame (n-frame) is a local geographic frame with its center at the system's location and its axes aligned with the local North, East, and Down directions. The n-frame rotates with respect to the e-frame at a rate defined by the

Pedestrian Inertial Navigation with Self-Contained Aiding, First Edition. Yusheng Wang and Andrei M. Shkel.
© 2021 The Institute of Electrical and Electronics Engineers, Inc. Published 2021 by John Wiley & Sons, Inc.

speed of the system relative to the Earth, since the Earth is spherical and the local N, E, and D directions will change. It is called transport rate, and it will be discussed further in Section 3.3. Note that geographic reference singularities exist over the north and south poles of the Earth in the n-frame.

- Wander azimuth frame (w-frame) is used to avoid the singularities in the polar regions and achieve world-wide navigation capability. Similar to the n-frame, the w-frame is also a locally leveled frame with its center at the system's location, but there is an azimuth angle between true north and the x-axis to make sure that the z-component of the transport rate is zero.
- Body frame (b-frame) is a frame fixed on the system to be measured. The three axes pointing to the front, right, and down directions can also be called the roll, pitch, and yaw axes.

3.2 Navigation Mechanism in the Inertial Frame

Inertial frame is the only frame mentioned above in which Newton's law of motion can be directly applied. Therefore, we start with the navigation in the i-frame.

In the i-frame, since it is non-rotational and non-accelerating, the accelerometer readout includes two parts: the actual acceleration of the system and the gravity, i.e.

$$a = f + g, \tag{3.1}$$

where a is the acceleration of the system, g is the gravitational acceleration, f is the accelerometer output and it is also called the specific force.

In most inertial navigation applications, Inertial Measurement Units (IMUs) are strapped to the system, and as a result, the accelerometer output is expressed in the b-frame. To perform navigation in the i-frame, coordinate transformation from the b-frame to the i-frame is needed:

$$f^i = C_b^i f^b, \tag{3.2}$$

where C_b^i is a 3×3 matrix defining the relative orientation of the b-frame with respect to the i-frame. It is called the Direction Cosine Matrix (DCM). The subscript denotes the reference frame in which the vector is expressed. The propagation of the DCM can be calculated from the angular rate measurements given by the gyroscopes:

$$\dot{C}_b^i = C_b^i [\omega_{ib}^b \times], \tag{3.3}$$

where ω_{ib}^b is the angular rate of the b-frame with respect to the i-frame, expressed in the b-frame, i.e. the gyroscope readout, and $[\cdot \times]$ is the skew-symmetric cross-product-operator. In other words, if $\omega_{ib}^b = [a \ b \ c]^T$, then

$$\Omega_{ib}^b \triangleq [\omega_{ib}^b \times] = \begin{bmatrix} 0 & -c & b \\ c & 0 & -a \\ -b & a & 0 \end{bmatrix}. \tag{3.4}$$

In most navigation applications, the quantities of interest are those relative to the e-frame instead of the i-frame. For example, the displacement and velocity with respect to the Earth make more sense than those with respect to some fixed stars that are billions of miles away. However, the inertial sensors measure the motion with respect to the i-frame. Therefore, a transformation between the e-frame and the i-frame is necessary. Coriolis' theorem is utilized to conduct the time derivative in the e-frame

$$\left. \frac{d\boldsymbol{r}_{ib}}{dt} \right|_i = \left. \frac{d\boldsymbol{r}_{ib}}{dt} \right|_e + \boldsymbol{\omega}_{ie} \times \boldsymbol{r}_{ib}, \tag{3.5}$$

where $\boldsymbol{r}_{ib}$ is the displacement of the b-frame with respect to the i-frame, and $\boldsymbol{\omega}_{ie}$ is the angular rate of the e-frame with respect to the i-frame, i.e. the Earth's rotation rate. Denote $\boldsymbol{v}_e = \left. \frac{d\boldsymbol{r}_{ib}}{dt} \right|_e$. If we take the time derivative of (3.5), then the acceleration can be expressed as

$$\boldsymbol{a} = \left. \frac{d^2 \boldsymbol{r}_{ib}}{dt^2} \right|_i = \left. \frac{d\boldsymbol{v}_e}{dt} \right|_i + \left. \frac{d}{dt}(\boldsymbol{\omega}_{ie} \times \boldsymbol{r}_{ib}) \right|_i = \left. \frac{d\boldsymbol{v}_e}{dt} \right|_i + \boldsymbol{\omega}_{ie} \times \boldsymbol{v}_e + \boldsymbol{\omega}_{ie} \times (\boldsymbol{\omega}_{ie} \times \boldsymbol{r}_{ib}). \tag{3.6}$$

Combining (3.1) and (3.6), we obtain the following navigation equation expressed in the i-frame

$$\left. \frac{d\boldsymbol{v}_e}{dt} \right|_i = C_b^i \boldsymbol{f}^b - \boldsymbol{\omega}_{ie}^i \times \boldsymbol{v}_e^i - \boldsymbol{\omega}_{ie}^i \times (\boldsymbol{\omega}_{ie}^i \times \boldsymbol{r}_{ib}^i) + \boldsymbol{g}^i. \tag{3.7}$$

Propagation of the DCM in the e-frame is similar to that in the i-frame except using the angular velocity with respect to the e-frame:

$$\dot{C}_b^e = C_b^e [\omega_{eb}^b \times], \tag{3.8}$$

where $\omega_{eb}^b = \omega_{ib}^b - C_e^b \omega_{ie}^e$ is the relative angular velocity of the b-frame with respect to the e-frame. Note that the Earth's rotation has been taken into consideration.

In (3.7), on the left-hand side is the acceleration of the system with respect to the Earth (denoted by the subscript) expressed in the i-frame. On the right-hand side, the first term is the specific force rotated by the DCM from the b-frame to the i-frame; the second term is the Coriolis acceleration caused by the rotation of the Earth; the third term is the centripetal force caused by the rotation of the Earth, and this part is often non-distinguishable from the gravitational acceleration $\boldsymbol{g}$. These two terms combined are called the local gravity vector $\boldsymbol{g}_l$ and it can be approximated by geodetic models, such as [4].

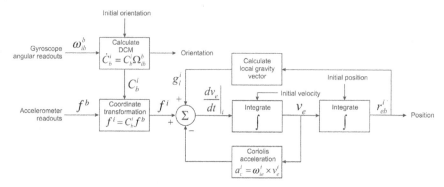

Figure 3.1 Block diagram of strapdown inertial navigation mechanism in the i-frame.

A block diagram of the strapdown inertial navigation mechanism in the i-frame is shown in Figure 3.1.

3.3 Navigation Mechanism in the Navigation Frame

In the n-frame, navigation data are expressed by velocity components along the North, East, and Down directions, and latitude, longitude, and altitude. Therefore, it is more commonly used in navigation applications on the Earth or in the vicinity of the Earth surface.

For orientation propagation, instead of transformation between the i-frame and the b-frame, we focus on the relation between the n-frame and the b-frame. The propagation of the DCM is similar to (3.3)

$$\dot{C}_b^n = C_b^n [\omega_{nb}^b \times], \tag{3.9}$$

where ω_{nb}^b is the relative angular rate of the b-frame with respect to the n-frame, and it can be calculated as

$$\omega_{nb}^b = \omega_{ib}^b - C_n^b(\omega_{ie}^n + \omega_{en}^n), \tag{3.10}$$

where ω_{en}^n is the relative angular rate of the n-frame with respect to the e-frame expressed in the n-frame, and it corresponds to the transport rate mentioned earlier.

For velocity propagation, similar to (3.5), changing rate of v_e in the n-frame can be written as

$$\frac{dv_e}{dt}\bigg|_n = \frac{dv_e}{dt}\bigg|_i - \omega_{in} \times v_e = \frac{dv_e}{dt}\bigg|_i - (\omega_{ie} + \omega_{en}) \times v_e. \tag{3.11}$$

Combining (3.1), (3.6), and (3.11), we will obtain

$$\frac{dv_e}{dt}\bigg|_n = a - (2\omega_{ie} + \omega_{en}) \times v_e - \omega_{ie} \times (\omega_{ie} \times r_{ib}). \tag{3.12}$$

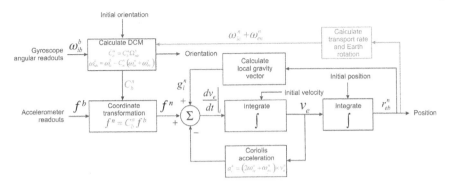

Figure 3.2 Block diagram of strapdown inertial navigation mechanism in the n-frame.

Equation (3.12) can be expressed in the n-frame coordinates as

$$\dot{v}_e^n = C_b^n f^b - (2\omega_{ie}^n + \omega_{en}^n) \times v_e^n - \omega_{ie}^n \times (\omega_{ie}^n \times r_{ib}^n) + g^n. \tag{3.13}$$

Note that in the n-frame inertial navigation mechanism, there are two terms in the Coriolis acceleration. The first term is due to the rotation of the Earth, and the second term can be considered a centripetal acceleration due to the motion on the surface of the Earth, which is assumed to be round. The Coriolis acceleration can be neglected in cases where the navigation error caused by the IMU measurement error is much greater than the Coriolis effect. These cases generally require a short navigation time (less than 10 minutes) and moderate IMU performances (tactical grade or worse).

A block diagram of the strapdown inertial navigation mechanism in the n-frame is shown in Figure 3.2. The differences between the mechanisms in the i-frame and the n-frame are shown in gray.

3.4 Initialization

As we discussed previously, inertial navigation equation uses the previous navigation solution as the starting point of current time step. Therefore, the system state, such as position, velocity, and attitude, needs to be initialized before navigation. Initialization of position and velocity cannot be achieved by inertial sensors, and requires external information, such as Global Navigation Satellite System (GNSS). In most cases, initialization is conducted when the vehicle is stationary with respect to the Earth. The position can be initialized at a known position obtained by the GNSS, and the velocity can be set to zero. Some residual motion during the initialization process may cause errors, and a common method to mitigate the errors is to perform initialization for longer time to average out the motion.

Attitude initialization, unlike position and velocity initialization, can be achieved by inertial sensors if the IMU is stationary with respect to the Earth. The roll and pitch angle can be extracted by measuring the direction of gravity by accelerometers. This process is called tilt sensing. After obtaining the roll and pitch angle, the yaw angle can be extracted by measuring the Earth's rotation using gyroscopes. This is called gyrocompassing. Another commonly used method to determine the yaw angle is by using triaxial magnetometer to measure the Earth's magnetic field. All three mentioned methods will be introduced in this section.

3.4.1 Tilt Sensing

Specific force measured by accelerometers contains two parts: linear acceleration and local gravity. The linear acceleration is typically zero during the initialization process, and therefore, the local gravity can be measured. The specific force can be derived from (3.13) with $\boldsymbol{v}_e^n$ set to be zero and the centripetal acceleration combined into the gravity

$$\boldsymbol{f}^b = -C_n^b \boldsymbol{g}^n, \tag{3.14}$$

where $\boldsymbol{g}^n$ is the local gravity vector. In the *zyx* intrinsic rotation convention, the DCM can be expressed as

$$C_n^b = \begin{bmatrix} \cos\theta\cos\psi & \cos\theta\sin\psi & -\sin\theta \\ -\cos\phi\sin\psi + \sin\phi\sin\theta\cos\psi & \cos\phi\cos\psi + \sin\phi\sin\theta\sin\psi & \sin\phi\cos\theta \\ \sin\phi\sin\psi + \cos\phi\sin\theta\cos\psi & -\sin\phi\cos\psi + \cos\phi\sin\theta\sin\psi & \cos\phi\cos\theta \end{bmatrix}, \tag{3.15}$$

where ϕ is the roll angle, θ is the pitch angle, and ψ is the yaw angle. Assuming the local gravity is along the Down direction in the n-frame with a magnitude of g, the specific force when the IMU is stationary can be expressed as:

$$\boldsymbol{f}^b = \begin{bmatrix} g\sin\theta \\ -g\sin\phi\cos\theta \\ -g\cos\phi\cos\theta \end{bmatrix}. \tag{3.16}$$

Thus, the roll and pitch angle can be estimated as

$$\phi = \arctan 2(-f_y^b, -f_z^b),$$

$$\theta = \arctan \frac{f_x^b}{\sqrt{f_y^{b2} + f_z^{b2}}}. \tag{3.17}$$

Note that the four-quadrant inverse tangent function is needed to determine the roll angle to reach a value in the closed interval $[-\pi, \pi]$. Tilt sensing is robust to accelerometer biases. Roll and pitch angle error of less than 0.1° can be achieved if the accelerometer bias is smaller than 2 mg, which is a reasonable requirement even for consumer grade IMUs. Yaw angle cannot be determined by tilt sensing since the local gravity is assumed to be along the Down direction and any rotation along the Down direction will not affect the specific force experienced by the IMU. More details of tilt sensing can be found in [5].

3.4.2 Gyrocompassing

When the IMU is stationary with respect to the n-frame, the rotation that it experiences is the Earth's rotation, which is along the z-axis of the e-frame. The yaw angle of the IMU can be determined by measuring the rotation in the b-frame except near the north and south poles.

When the IMU is stationary, (3.10) can be reduced to

$$
\begin{bmatrix} \omega_x^b \\ \omega_y^b \\ \omega_z^b \end{bmatrix} = \omega_{ib}^b = C_n^b \omega_{ie}^n = C_n^b \begin{bmatrix} \Omega \cos L \\ 0 \\ -\Omega \sin L \end{bmatrix},
\tag{3.18}
$$

where Ω is the magnitude of the Earth's rotation and L is the latitude of the IMU. Combining (3.15) and (3.18), and assuming that roll and pitch angles are available from tilt sensing, we can determine the yaw angle of the IMU without knowing the latitude and the Earth's rotation rate [6]:

$$
\psi = \arctan 2(s, c),
\tag{3.19}
$$

where

$$
s = \omega_z^b \sin \phi - \omega_y^b \cos \phi,
$$
$$
c = \omega_x^b \cos \theta + \omega_y^b \sin \phi \sin \theta + \omega_z^b \cos \phi \sin \theta.
$$

Note that the four-quadrant inverse tangent function is needed.

Gyrocompassing requires the IMU to measure the Earth's rotation, which is around 15 °/h, to obtain the yaw angle information. Typically, only navigation grade IMUs can achieve the goal, and long averaging time may be required to compensate for sensor noises and potential vibrations. For example, it was reported in [7] that 0.23° of yaw angle estimation uncertainty was demonstrated with the gyroscope of 0.2 °/h bias instability under continuous carouseling.

The relation between the gyroscope bias and yaw angle estimation error is discussed in [1], and the result is reproduced in Figure 3.3. In this model, the IMU was assumed to be stationary with respect to the ground and approximately aligned to the NED axes of the n-frame. Note that the yaw angle error is not only related to the gyro bias but also to the latitude. The estimation error is much larger in the area close to the north pole. In most areas on the Earth, a *constant* gyroscope bias of

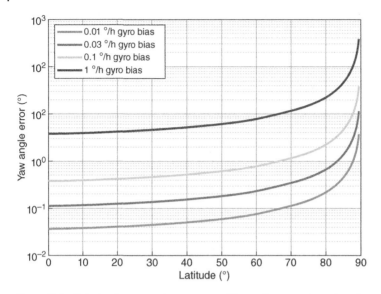

Figure 3.3 Relation between the gyroscope bias and yaw angle estimation error. Source: Modified from Titterton and Weston [1].

less than 0.03 °/h (not bias instability) is needed to achieve gyrocompassing with less than 0.1° of yaw angle estimation error.

3.4.3 Magnetic Heading Estimation

High-performance requirements are posed on the gyroscopes in the gyrocompassing to determine the yaw angle. Therefore, implementation of external information, such as the Earth's magnetic field, is necessary in most applications where tactical or consumer grade IMUs are used.

In the magnetic heading estimation, a triaxial magnetometer is needed to measure the direction and the magnitude of the Earth's magnetic field. The principle is similar to that of gyrocompassing, except that the Earth's magnetic field is measured instead of the Earth's rotation. Therefore, similar equations to determine the yaw angle are used:

$$\psi' = \arctan 2(s, c), \tag{3.20}$$

where

$$\begin{aligned} s &= m_z^b \sin \phi - m_y^b \cos \phi, \\ c &= m_x^b \cos \theta + m_y^b \sin \phi \sin \theta + m_z^b \cos \phi \sin \theta, \end{aligned} \tag{3.21}$$

where m_x^b, m_y^b, and m_z^b are the three components of magnetometer measurement along x, y, and z axis of the b-frame, respectively. Note that the true yaw angle can be calculated as

$$\psi = \psi' + \alpha, \tag{3.22}$$

where α is the declination angle of the Earth's magnetic field. This is due to the misalignment of the axis of the Earth's magnetic field and the Earth's axis of rotation. The declination angle varies in a predictable way, and it can be found in the World Magnetic Model [8].

The limitations of magnetic heading estimation are also obvious: the Earth's magnetic field may be distorted or affected by many factors, such as equipment installed on the host vehicle, hard-iron and soft-iron magnetism, and local magnetic anomalies. A few degrees of estimate bias is common in a complex electromagnetic environment. However, compensation methods based on extended Kalman filter have been demonstrated to mitigate the estimation errors [9].

3.5 Conclusions

In this chapter, fundamentals of strapdown inertial navigation mechanism were briefly introduced. More specifically, algorithms to integrate IMU readouts into navigation state information and techniques to conduct initialization were discussed. Since the navigation result is highly nonlinear with respect to the IMU readouts, the navigation error accumulation is also complicated. As a result, it is necessary to analyze the relation between the IMU error and the navigation error, in order to be able to select the proper IMUs for certain navigation applications. This leads to the discussion of Chapter 4.

References

1 Titterton, D. and Weston, J. (2004). *Strapdown Inertial Navigation Technology*, 2e, vol. 207. AIAA.

2 Britting, K.R. (1971). *Inertial Navigation Systems Analysis*. Wiley.

3 Savage, P.G. (2007). *Strapdown Analytics*, 2e. Maple Plain, MN: Strapdown Associates.

4 Stieler, B. and Winter, H. (1982). *Gyroscopic Instruments and Their Application to Flight Testing, AGARD Flight Test Instrumentation Series*, vol. 15, No. AGARD-AG-160-VOL-15. Advisory Group for Aerospace Research and Development Neuilly-sur-Seine (FRANCE).

5 Pedley, M. (2012–2013). Tilt sensing using a three-axis accelerometer. *Freescale semiconductor application note.*

6 Groves, P.D. (2015). Navigation using inertial sensors [Tutorial]. *IEEE Aerospace and Electronic Systems Magazine* 30 (2): 42–69.

7 Prikhodko, I.P., Zotov, S.A., Trusov, A.A., and Shkel, A.M. (2013). What is MEMS gyrocompassing? Comparative analysis of maytagging and carouseling. *Journal of Microelectromechanical Systems* 22 (6): 1257–1266.

8 Chulliat, A., Macmillan, S., Alken, P. et al. (2015). The US/UK World Magnetic Model for 2015–2020.

9 Guo, P., Qiu, H., Yang, Y., and Ren, Z. (2008). The soft iron and hard iron calibration method using extended Kalman filter for attitude and heading reference system. *2008 IEEE/ION Position, Location and Navigation Symposium (PLANS)*, Monterey, CA, USA (5–8 May 2008).

4

Navigation Error Analysis in Strapdown Inertial Navigation

One of the most important characteristics of an inertial navigation system is its navigation accuracy, which is directly related to the measurement errors of the Inertial Measurement Unit (IMU). However, a number of questions need to be answered: How to characterize these errors? How are these errors related to each other? What are the requirements on the IMU measurement accuracy according to required navigation accuracy? This chapter intends to answer these questions.

In this chapter, we analyze the relation between the IMU error and the navigation error in the strapdown inertial navigation. First, we introduce the terminology, origin, and characteristics of some of the major error sources. It is followed by IMU calibration techniques to remove the constant part of the errors. Then, navigation error accumulation is derived and analyzed.

4.1 Error Source Analysis

Error sources in navigation can be categorized into three groups:

- *IMU errors.* An IMU is typically an assembly of three accelerometers and three gyroscopes mounted perpendicular to each other. As a result, the IMU error is composed of two parts: the error from individual inertial sensors and the error arising from the assembly process. Typical inertial sensor errors include bias, random noise, scale factor error, error caused by shock and vibration, temperature drift, etc. Assembly errors are mainly related to the misalignment of inertial sensors to the frame, and they are also one of the main error sources in the navigation system. Since IMU errors are inevitable and sometimes the dominant ones during navigation, our main focus will be on IMU errors in this chapter.

Pedestrian Inertial Navigation with Self-Contained Aiding, First Edition. Yusheng Wang and Andrei M. Shkel.
© 2021 The Institute of Electrical and Electronics Engineers, Inc. Published 2021 by John Wiley & Sons, Inc.

- *Initial calibration errors.* This category is related to the errors introduced by initialization of the system. The common sources are inaccuracy of initial position, velocity, and orientation information, and misalignment of the sensors. A careful initial calibration of the system can greatly reduce errors of this kind. This part has been briefly introduced in Chapter 2, and will not be covered in this chapter.
- *Numerical errors.* This category is related to the errors created by numerical computations, such as round-off errors and approximation of derivatives by finite differences. Errors of this kind can be controlled with properly designed algorithms.

4.1.1 Inertial Sensor Errors

Common inertial sensor errors are listed in Figure 4.1. The dashed lines are the ideal case, and the solid lines represent the output with different sources of noise. Note that in Figure 4.1a, the noise is stochastic and the solid line does not represent the input–output relation. In Figure 4.1b–f, the errors are all deterministic. In this section, we only focus on the noise, bias, and scale factor errors of the inertial sensors. The effects of nonlinearity, dead zone, and quantization will not be dominant as long as a proper measuring range and resolution of the sensor are selected.

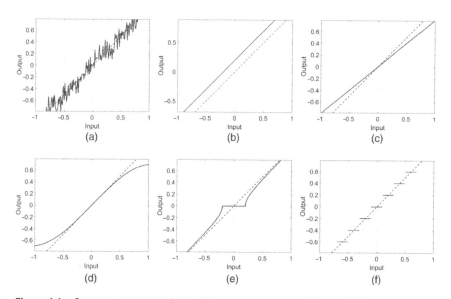

Figure 4.1 Common error types in inertial sensor readouts. (a) Noise, (b) bias, (c) scale factor error, (d) nonlinearity, (e) dead zone, (f) quantization.

Bias and scale factor errors are relatively constant compared to noises in the inertial sensor measurements. Therefore, it is easier to calibrate and compensate for them before conducting the navigation. Noises in inertial sensors come from different origins and they show different characteristics. Some of the main noise types are briefly introduced in this section. Further material can be found in [1].

- *Quantization noise.* It is one of the noises associated with Analog-to-Digital Conversion (ADC). It is caused by small differences between the actual amplitudes of the analog signal being sampled and the bit resolution of the analog-to-digital converter. It is a noise with an f^2 Power Spectral Density (PSD) and it shows up on Allan Deviation (AD) [2] with a slope of τ^{-1} (see Figure 4.2).
- *Angle (Velocity) Random Walk.* This kind of noise is typically caused by white thermomechanical and thermoelectrical noise with much higher frequency than the sampling rate. It is characterized by a white-noise spectrum, meaning it has equal intensity at different frequencies. It shows a slope of $\tau^{-0.5}$ on AD. The effect of this kind of noise is the random walk of angle estimation (in the case of gyroscopes) and velocity estimation (in the case of accelerometers) during navigation.
- *Bias instability.* The origin of this noise is usually due to the electronics susceptible to random flickering. It is indicated as a sensor bias fluctuation in the data. It has a $1/f$ PSD under the cutoff frequency, and it shows a curve on AD with a slope of zero.
- *Rate (accelerometer) random walk.* The origin of this noise is typically unknown. It has a $1/f^2$ PSD and shows as a curve on AD with a slope of $\tau^{+0.5}$. The effect can be considered white noise applied on the bias of the sensor. Therefore, it is also called "random walk."
- *Drift rate ramp.* This kind of error is related to output change due to temperature change. Strictly speaking, it is not a stochastic error, but we still can analyze it in a similar manner. It has a $1/f^3$ PSD and shows a curve on AD with a slope of τ^{+1}.

A simplified model to describe the inertial sensor errors can be written as [3]

$$\text{output}_i = \text{input}_i + c_i + b_i + w_i, \tag{4.1}$$

where the subscript i can be either gyroscope or accelerometer, c is the constant offset, b is the walking bias, and w is the wide band sensor noise.

The constant offset might be the result of turn-on bias, and it will remain constant as long as the sensor is on. Therefore, this part can be easily estimated during calibration before conducting navigation. The walking bias term is also called sensor drift, and it is often modeled as first order Gauss–Markov model, which can be

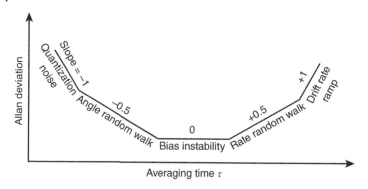

Figure 4.2 A schematic of log–log plot of Allan deviation. Source: Modified from IEEE Std [2].

determined by its Standard Deviation (SD) σ_{bias} and time constant τ_c

$$b_{k+1} = \left(1 - \frac{\Delta t}{\tau_c}\right) b_k + \sqrt{\frac{2\sigma_{bias}^2}{\tau_c}} \Delta t \cdot v_k, \tag{4.2}$$

where Δt is the length of each time step, and v_k is a sequence of standard Gaussian distribution. Notice that an infinite time constant is assumed for the Gauss–Markov model for sensor bias propagation. In practice, the time constant τ_c is generally on the order of 10 to 100 seconds for MEMS-based (Micro-Electro-Mechanical Systems) accelerometers and gyroscopes, and the IMU sampling frequency is generally above 100 Hz. Therefore, the time constant is at least three to four magnitudes larger than Δt. As a result, the walking bias can be approximated as the accumulation of a white noise sequence of Gaussian distribution.

The wide band sensor noise w is generally modeled to be normally distributed with a zero mean and sampled covariance of

$$E[w_2] = \sigma^2 \cdot f, \tag{4.3}$$

where f is the sampling frequency. The AD of the wide band sensor noise has a slope of $\tau^{-0.5}$, and it corresponds to the Angle Random Walk (ARW) (in the case of gyroscopes) or the Velocity Random Walk (VRW) (in the case of accelerometers). The AD of the walking bias has a slope of $\tau^{0.5}$ if $\tau \ll \tau_c$, corresponding to the Rate Random Walk (RRW) (in the case of gyroscopes) or Accelerometer Random Walk (AcRW) (in the case of accelerometers).

The g-sensitivity of gyroscopes is another kind of inertial sensor errors, which is the erroneous measurement of a gyroscope in response to the external acceleration. The g-sensitivity of gyroscopes is typically the result of a few factors: deformation of the device due to external acceleration, asymmetric structure of the device, and the coupling between different resonant modes within the device. Therefore, the gyroscope g-sensitivity is most observed in MEMS-based gyroscopes. On the contrary, optical gyroscopes, such as ring laser gyroscopes, are generally immune to it. The magnitude of the g-sensitivity of a gyroscope can be characterized by the g-sensitivity matrix [4]

$$\begin{bmatrix} \omega_x \\ \omega_y \\ \omega_z \end{bmatrix} = \begin{bmatrix} g_{xx} & g_{xy} & g_{xz} \\ g_{yx} & g_{yy} & g_{yz} \\ g_{zx} & g_{zy} & g_{zz} \end{bmatrix} \begin{bmatrix} a_x \\ a_y \\ a_z \end{bmatrix}, \tag{4.4}$$

where a_x, a_y, and a_z are the components of specific force that the IMU experiences along x, y, and z directions, respectively, and ω_x, ω_y, and ω_z are the corresponding readout due to g-sensitivity. The on-diagonal terms are usually one order of magnitude higher than the off-diagonal terms, and thus, in cases where the gyroscope g-sensitivity is not dominant, the g-sensitivity matrix can be simplified as diagonal a matrix. Notice that the g-sensitivity matrix does not have to be symmetric.

4.1.2 Assembly Errors

Assembly errors are another major source of IMU errors. Assembly errors mainly come from the shift of the mounting direction of individual inertial sensors from their ideal orientations. A schematic of the IMU assembly error is shown in Figure 4.3, where the black arrows are the three ideally orthogonal direction *xyz*, and the light gray arrows are the actual orientations of the mounted inertial sensors *x'y'z'*, and the orientation of each sensor can be characterized by two angles. The assembly error of inertial sensors in the IMU will cause not only deviation of the effective scale factor along certain directions but also the cross-talk between axes due to the coupling of different directions.

The assembly error can be decomposed of two parts: non-orthogonality and misalignment. The non-orthogonality describes how orientation of each

Figure 4.3 A schematic of the IMU assembly error.

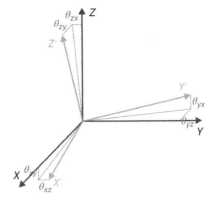

individual sensor is off from orthogonality, which is mainly induced during the IMU assembly process. The misalignment describes how the accelerometer or gyroscope triad as one body is off from the ideal orientation, which is caused during the mounting of the IMU to the navigation system. To better illustrate the concepts of non-orthogonality and misalignment, an intermediate reference frame $\hat{x}\hat{y}\hat{z}$ is introduced, whose three axes are orthogonal to each other, and $\hat{x}$ axis is aligned with the x axis of the triad (see Figure 4.4). Then, the non-orthogonality is presented by the difference between the inertial sensor triad $x'y'z'$ and the frame $\hat{x}\hat{y}\hat{z}$, whereas the misalignment is the difference between the frame $\hat{x}\hat{y}\hat{z}$ and xyz. As for the non-orthogonality, the orientation difference between the y' and $\hat{y}$ can be defined using one parameter α_{yz}, with the subscript indicating the rotation of the y-axis with respect to the z-axis. Next, the non-orthogonality of the $\hat{z}$ axis can be expressed by a rotation around the x-axis by α_{zx} followed by a rotation around the y-axis by α_{zy}. All the notations are presented in Figure 4.4. The non-orthogonal axes $x'y'z'$ can be presented as

$$
\begin{bmatrix} x' \\ y' \\ z' \end{bmatrix} = \begin{bmatrix} 1 & 0 & 0 \\ -\sin\alpha_{yz} & \cos\alpha_{yz} & 0 \\ \sin\alpha_{zy} & -\sin\alpha_{zx}\cos\alpha_{zy} & \cos\alpha_{zx}\cos\alpha_{zy} \end{bmatrix} \begin{bmatrix} \hat{x} \\ \hat{y} \\ \hat{y} \end{bmatrix}
$$
$$
\approx \begin{bmatrix} 1 & 0 & 0 \\ -\alpha_{yz} & 1 & 0 \\ \alpha_{zy} & -\alpha_{zx} & 1 \end{bmatrix} \begin{bmatrix} \hat{x} \\ \hat{y} \\ \hat{y} \end{bmatrix},
\tag{4.5}
$$

where the approximation holds if the non-orthogonal angles are small. The misalignment can be described simply by a rotation transformation, which can be presented by a Direction Cosine Matrix (DCM).

Combining the non-orthogonality, misalignment, and the scale factor errors of the IMU, the input–output relation of an IMU (neglecting other types of error) can

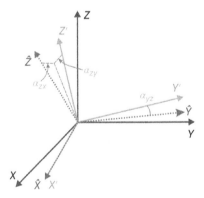

Figure 4.4 Illustration of the two components of the IMU assembly error: non-orthogonality and misalignment.

be expressed as

$$
\begin{bmatrix} \text{output}_x \\ \text{output}_y \\ \text{output}_z \end{bmatrix} = \begin{bmatrix} 1+s_x & m_{xy} & m_{xz} \\ m_{yx} & 1+s_y & m_{yz} \\ m_{zx} & m_{zy} & 1+s_z \end{bmatrix} \begin{bmatrix} \text{input}_x \\ \text{input}_y \\ \text{input}_z \end{bmatrix}, \tag{4.6}
$$

where s_x, s_y, and s_z are the scale factor errors along the three nominal directions, and m's are the parameters describing the misalignment. If the misalignment angles are small, the following approximation relations exist

$$
\begin{aligned}
m_{yx} &\approx -\theta_{yz}, & m_{zx} &\approx \theta_{zy}, & m_{xy} &\approx \theta_{xz}, \\
m_{zy} &\approx -\theta_{zx}, & m_{xz} &\approx -\theta_{xy}, & m_{yz} &\approx \theta_{yx}.
\end{aligned} \tag{4.7}
$$

To sum up, a full model describing the IMU errors can be expressed as [5]

$$
\begin{aligned}
\begin{bmatrix} y_{Ax} \\ y_{Ay} \\ y_{Az} \end{bmatrix} &= \begin{bmatrix} 1+s_{Ax} & m_{Axy} & m_{Axz} \\ m_{Ayx} & 1+s_{Ay} & m_{Ayz} \\ m_{Azx} & m_{Azy} & 1+s_{Az} \end{bmatrix} \begin{bmatrix} A_x \\ A_y \\ A_z \end{bmatrix} \\
&\quad + \begin{bmatrix} c_{Ax} \\ c_{Ay} \\ c_{Az} \end{bmatrix} + \begin{bmatrix} b_{Ax} \\ b_{Ay} \\ b_{Az} \end{bmatrix} + \begin{bmatrix} w_{Ax} \\ w_{Ay} \\ w_{Az} \end{bmatrix}, \\
\begin{bmatrix} y_{Gx} \\ y_{Gy} \\ y_{Gz} \end{bmatrix} &= \begin{bmatrix} 1+s_{Gx} & m_{Gxy} & m_{Gxz} \\ m_{Gyx} & 1+s_{Gy} & m_{Gyz} \\ m_{Gzx} & m_{Gzy} & 1+s_{Gz} \end{bmatrix} \begin{bmatrix} G_x \\ G_y \\ G_z \end{bmatrix} \\
&\quad + \begin{bmatrix} g_{xx} & g_{xy} & g_{xz} \\ g_{yx} & g_{yy} & g_{yz} \\ g_{zx} & g_{zy} & g_{zz} \end{bmatrix} \begin{bmatrix} A_x \\ A_y \\ A_z \end{bmatrix} + \begin{bmatrix} c_{Gx} \\ c_{Gy} \\ c_{Gz} \end{bmatrix} + \begin{bmatrix} b_{Gx} \\ b_{Gy} \\ b_{Gz} \end{bmatrix} + \begin{bmatrix} w_{Gx} \\ w_{Gy} \\ w_{Gz} \end{bmatrix},
\end{aligned} \tag{4.8}
$$

where the subscript G denotes gyroscope, the subscript A denotes accelerometer, and the rest symbols are consistent with previous notations.

Notice that in most navigation applications, the total navigation time is less than a few hours, and the requirement of the navigation accuracy is not extremely high. Thus, the scale factor error, IMU misalignment, and the g-sensitivity may be considered constant throughout the navigation. As a result, they can be calibrated and compensated before the navigation, and their effects on the final navigation error may be compensated [6]. The calibration method will be discussed later in this chapter.

4.1.3 Definition of IMU Grades

IMUs are categorized by different grades according to their performance characteristics. There is no specific definition on the grades, but IMUs can still be generally categorized into four grades: consumer grade, industrial grade, tactical

Table 4.1 Classification of IMU performances in terms of bias instability.

	Accelerometer BI (mg)	Gyroscope BI (°/h)	Typical applicable field
Consumer grade	>50	>100	Consumer electronics
Industrial grade	1–50	10–100	Automotive industry
Tactical grade	0.05–1	0.1–10	Short-term navigation
Navigation grade	<0.05	<0.01	Aeronautics navigation

Source: Based on Passaro et al. [8], Yazdi et al. [9].

grade, and navigation grade. One of the common standard for categorization is the Bias Instability (BI) of the sensors [7], and it can be measured as the minimum on the Allan Variance curve. A typical classification of IMUs is listed in Table 4.1.

4.1.3.1 Consumer Grade
The lowest grade of inertial sensors are often referred to as consumer grade. Due to their high noise level, consumer grade inertial sensors are typically sold as individual accelerometers or gyroscopes, instead of in the form of a complete IMU. However, this is changing as more high-performance MEMS-based devices are appearing in the market. The position error will exceed a meter of error within a few seconds of navigation with unaided consumer grade IMUs. Consumer grade inertial sensors are typically used in consumer electronics, such as smart phones, tablets, gaming controllers, and entertainment systems. Most consumer grade inertial sensors are made using lithography-based MEMS technologies due to the low cost resulted from batch fabrication.

4.1.3.2 Industrial Grade
Industrial grade inertial sensors generally have similar or better noise performances than the consumer grade inertial sensors, but with a better calibration procedure. Due to their relatively low performance, industrial grade IMUs are typically implemented with aiding from other sensors, such as magnetometer and barometer, to be used in navigation-related applications. Estimation methods, such as Extended Kalman Filter (EKF), are typically used to compute the result from multiple sources. Typical applications of industrial grade IMUs include Attitude and Heading Reference System (AHRS), automotive applications, such as Anti-lock Braking System (ABS), active suspension system, airbag, and aided pedestrian dead-reckoning system. Industrial grade inertial sensors are generally made using MEMS technologies.

4.1.3.3 Tactical Grade

Tactical grade IMUs have the capability of attitude measurement with reasonable errors and are able to conduct short-term navigation, with navigation accuracy on the order of meters within 30 seconds of strapdown inertial navigation. Navigation accuracy on the order of centimeters can be achieved with the integration with GPS [10]. Tactical grade inertial sensors can be made from MEMS, fiber optic gyroscope (FOG), and ring laser gyroscope (RLG) technologies.

4.1.3.4 Navigation Grade

Navigation grade IMUs are some of the highest performance devices that are available for general purpose applications and are applicable for aeronautical navigation with navigation error better than 1 nautical mile per hour. The devices with performance higher than the Navigation grade (Navigation +) are also available, but are only considered for a very specialized type of applications, such as navigation for submarines. Such a high-end IMU can cost over $1 million, and they can typically provide navigation solution drifts less than 1 nautical mile per day without any aiding. Navigation grade inertial sensors and above are made from RLG and precision machining technologies.

Table 4.2 lists some of the commercially available IMUs and their characteristics.

4.2 IMU Error Reduction

4.2.1 Six-Position Calibration

In this section, we discuss the IMU calibration techniques, so that the relatively time-invariant IMU errors, such as the constant offset, scale factor errors, non-orthogonality, and the gyroscope g-sensitivity can be compensated.

The six-position static and rate test are the most commonly used techniques to calibrate accelerometers and gyroscopes in the IMU. This process requires one axis of the IMU to be aligned with the local navigation frame, pointing alternatively up and down. Thus, a total of six different positions are involved in the calibration process. If only the constant offset c and the scale factor error s are considered, the measurement result can be expressed as

$$
\begin{aligned}
y^{\mathrm{up}} &= (1 + s) \cdot x + c, \\
y^{\mathrm{down}} &= -(1 + s) \cdot x + c,
\end{aligned}
\tag{4.9}
$$

where y^{up} and y^{down} are the sensor measurement when the measuring axis is pointing upward and downward, respectively, and x is the actual measurement input,

Table 4.2 List of some commercial IMUs and their characteristics.

Company	Product name	ARW ($°/h^{0.5}$)	Gyro BI ($°/h$)	VRW ($µg/Hz^{0.5}$)	Accel BI ($µg$)	Volume	Grade
Bosch	BMI160	0.42	252[a]	180	1800[a]	6.0 mm³	Consumer
STMicroelectronics	ISM330DLC	0.228	270[a]	75	1800[a]	6.225 mm³	Consumer
TDK InvenSense	ICM-42605	0.228	136.8[a]	70	700[a]	6.825 mm³	Industrial
Analog Devices	ADI16495	0.12	0.8	13.6	3.2	29 cm³	Tactical
Honeywell	HG1930	<0.06	0.25	<51.8	20	82 cm³	Tactical
Systron Donner	SDI500	<0.02	1	100	100	310 cm³	Tactical
Northrop Grumman	LN-200S	<0.07	1	35	300	570 cm³	Tactical
Safran	PRIMUS 400	<0.002	<0.01	<60	<1000	520 cm³	Navigation
Honeywell	HG9900	<0.002	<0.003	—[b]	<10	1700 cm³	Navigation
GEM elettronica	IMU-3000	0.0008	0.002	—[b]	<40	—[b]	Navigation

a) Data calculated from total RMS noise.
b) Data not available.
Source: Based on Bosch, STMicroelectronics, InvenSense, Analog Devices, Honeywell, Systron Donner, Northrop Grumman, Safran, Honeywell, GEM elettronica [11–20].

which is the local gravity or the Earth's rotation rate in the case of accelerometers and gyroscopes, respectively. For low-cost MEMS IMUs whose noise level is on the same order or even higher than the Earth's rotation, a rate table can be used to generate a stable and constant rotation as the gyroscope input. The constant offset and scale factor error can be obtained as follows

$$c = \frac{1}{2}(y^{\text{up}} + y^{\text{down}}),$$

$$s = \frac{1}{2x}(y^{\text{up}} - y^{\text{down}}) - 1.$$

(4.10)

Notice that the actual measurement input x is needed to determine the scale factor error s. Therefore, an accurate external reference for both acceleration and rotation is needed. The full IMU error model including the non-orthogonality and the gyroscope g-sensitivity can also be incorporated in the six-position tests. The least squares method can be used in this case to derive all the parameters. Calibration accuracy of this method is dependent on how well the IMU is aligned with the local level frame. A perfect cubic-shaped mounting frame is typically used for better calibration accuracy [21].

4.2.2 Multi-position Calibration

To avoid the fine alignment of the IMU during calibration, another method called multi-position method is developed [22]. In this method, no precise alignment of the IMU to the gravity is needed, therefore making the whole process more convenient to be carried out in the field.

In the multi-position calibration method, the IMU is placed stationary with different orientations and the readouts are collected. Unlike the six-position calibration method, where the orientation is already defined, the orientation of the IMU can be selected randomly. However, at least nine different orientations should be measured to avoid singularity in the following data processing. The general idea during data processing is to fit the IMU error parameters such that the magnitude of the measured signal is equal to the external reference, which is the local gravity for accelerometer triad calibration, and the Earth's rotation rate for gyroscope triad calibration. The parameters can be obtained by iterative weighted least squares method [23]. The Earth's rotation rate can be replaced by a reference rotation created by a rate table, if the IMU noise level is too high to measure the Earth's rotation [21].

4.3 Error Accumulation Analysis

It has been mentioned previously that navigation errors will be accumulated in strapdown inertial navigation. But questions, such as "at what rate are the errors accumulated?" and "what is the main error source during the navigation?," have not been answered. In this section, we identify some of the major error sources in strapdown inertial navigation and introduce some error models based on the navigation mechanism in the n-frame.

4.3.1 Error Propagation in Two-Dimensional Navigation

In Section 4.1, we introduced some of the major errors in the inertial sensors, and in this section, we look into how the errors mentioned above propagate during the navigation and how they affect the final navigation results. We start with an error propagation model in a two-dimensional i-frame navigation in the x–z plane (Figure 4.5). For simplicity, Coriolis acceleration due to the Earth's rotation and transport rate are neglected. A constant local gravity vector is also assumed.

The equations describing the navigation mechanism are as follows

$$\dot{v}_x^i = f_x^i = f_x^b \cos\theta + f_z^b \sin\theta, \tag{4.11}$$

$$\dot{v}_z^i = f_z^i - g = -f_x^b \sin\theta + f_z^b \cos\theta - g, \tag{4.12}$$

$$\dot{\theta} = \omega_y^b, \qquad \dot{x}^i = v_x^i, \qquad \dot{z}^i = v_z^i. \tag{4.13}$$

The equations describing the error propagation can be derived by taking partial derivative to (4.11)–(4.13)

$$\delta\dot{v}_x^i = \delta f_x^b \cos\theta + \delta f_z^b \sin\theta - f_x^b \sin\theta\,\delta\theta + f_z^b \cos\theta\,\delta\theta$$
$$= \delta f_x^b \cos\theta + \delta f_z^b \sin\theta + f_z^i \delta\theta, \tag{4.14}$$

$$\delta\dot{v}_z^i = -\delta f_x^b \sin\theta + \delta f_z^b \cos\theta - f_x^b \sin\theta\,\delta\theta - f_z^b \cos\theta\,\delta\theta$$
$$= -\delta f_x^b \sin\theta + \delta f_z^b \cos\theta - f_x^i \delta\theta, \tag{4.15}$$

$$\delta\dot{\theta} = \delta\omega_y^b, \qquad \delta\dot{x}^i = \delta v_x^i, \qquad \delta\dot{z}^i = \delta v_z^i. \tag{4.16}$$

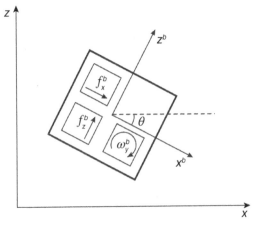

Figure 4.5 Two-dimensional strapdown inertial navigation system in a fixed frame. Two accelerometers and one gyroscope is needed.

Since it is a set of linear equations with respect to the error terms, we can analyze the different error sources separately, and the final navigation result is simply the summation of each term. We first study the effects of some common deterministic errors, for example, fixed gyroscope bias error $\delta\omega_y^b$. From the first equation in (4.16), we can calculate $\delta\theta = \delta\omega_y^b \cdot t$. Next, by taking $\delta\theta$ into (4.14), we will obtain $\delta v_x^i = \frac{1}{2}\delta\omega_y^b f_z^i \cdot t^2$. Finally, substituting δv_x^i in the second equation in (4.16), the position error along the x-axis can be expressed as

$$\delta x^i = \frac{1}{6}\delta\omega_y^b f_z^i \cdot t^3. \tag{4.17}$$

Similarly, the position error along the z-axis can be calculated

$$\delta z^i = -\frac{1}{6}\delta\omega_y^b f_x^i \cdot t^3. \tag{4.18}$$

The relation between the position estimation errors along two directions and deterministic errors is presented in Table 4.3. Note that in a real navigation problem, the errors mentioned above will couple to each other due to rotation of the body frame. Therefore, a rigorous error estimation is usually much more complicated than what is presented in the table. However, Table 4.3 shows how different errors propagate during navigation. For example, position error due to gyroscope bias propagates as t^3, whereas the position error due to accelerometer bias propagated as t^2, indicating that gyroscope biases are dominant in long-term navigation compared to accelerometer biases.

Next, we look into the effects of common stochastic errors. Due to the stochastic nature of the error source, there is no fixed relation between the navigation error

Table 4.3 Propagation of position errors in 2D strapdown inertial navigation due to deterministic errors.

		Position estimation error	
Deterministic error source		Along x-axis	Along z-axis
Initial position error	δx_0	δx_0	0
	δz_0	0	δz_0
Initial velocity error	δv_{x0}	$\delta v_{x0} \cdot t$	0
	δv_{z0}	0	$\delta v_{z0} \cdot t$
Initial orientation error	$\delta\theta_0$	$\frac{1}{2}\delta\theta_0 f_z^i \cdot t^2$	$-\frac{1}{2}\delta\theta_0 f_x^i \cdot t^2$
Accelerometer bias	δf_x^b	$\frac{1}{2}\delta f_x^b \cos\theta \cdot t^2$	$-\frac{1}{2}\delta f_x^b \sin\theta \cdot t^2$
	δf_z^b	$\frac{1}{2}\delta f_z^b \sin\theta \cdot t^2$	$\frac{1}{2}\delta f_z^b \cos\theta \cdot t^2$
Gyroscope bias	$\delta\omega_y^b$	$\frac{1}{6}\delta\omega_y^b f_z^i \cdot t^3$	$-\frac{1}{6}\delta\omega_y^b f_x^i \cdot t^3$

and the amplitude of the stochastic errors. However, the variance of the navigation error is determined by the amplitude of the error and can be studied. For example, in the case of ARW, the continuous sensor error signal $\epsilon(t)$ can be discretized as a sequence of zero-mean, identically distributed, uncorrelated random variables N_i with a finite variance σ^2 [24]. As a result, the integral of the error can be expressed as the summation of the sequence

$$I_1 = \int_0^t \epsilon(\tau)d\tau = \delta t \sum_{i=1}^n N_i, \tag{4.19}$$

where δt is the time between successive samples, and $t = n\delta t$. Then, we can obtain the mean and variance of the integral I_1:

$$E[I_1] = E\left[\delta t \sum_{i=1}^n N_i\right] = \delta t \sum_{i=1}^n E[N_i] = 0, \tag{4.20}$$

$$\text{Var}[I_1] = \text{Var}\left[\delta t \sum_{i=1}^n N_i\right] = \delta t^2 \sum_{i=1}^n \text{Var}[N_i] = \delta t \cdot t \cdot \sigma^2. \tag{4.21}$$

From the above two equations, we conclude that ARW introduces a zero-mean error in the angle estimation with a SD of

$$\sigma_\theta(t) = \sqrt{\text{Var}[I]} = \sigma \cdot \sqrt{\delta t \cdot t}. \tag{4.22}$$

Since ARW is defined as $\sigma_\theta(1) = \sigma \cdot \sqrt{\delta t}$, then

$$\sigma_\theta(t) = \text{ARW} \cdot \sqrt{t}. \tag{4.23}$$

Equation (4.23) shows that the SD of angle estimation in strapdown navigation due to ARW propagates as $t^{1/2}$. The error of position estimation along the x-axis can be expressed as

$$I_3 = \int_0^t \int_0^t f_z^i \int_0^t \epsilon(\tau)d\tau d\tau d\tau$$
$$= f_z^i \delta t^3 \sum_{i=1}^n \sum_{i=1}^n \sum_{i=1}^n N_i = f_z^i \delta t^3 \sum_{i=1}^n \frac{i(i+1)}{2} N_{n-i+1}. \tag{4.24}$$

Similarly, the mean and variance of the integral can be calculated as

$$E[I_3] = f_z^i \delta t^3 \sum_{i=1}^n E\left[\frac{i(i+1)}{2} N_{n-i+1}\right] = 0, \tag{4.25}$$

$$\text{Var}[I_3] = (f_z^i)^2 \delta t^6 \sum_{i=1}^n \left[\frac{i(i+1)}{2}\right]^2 \text{Var}[N_{n-i+1}] \approx \frac{1}{20}(f_z^i)^2 \cdot \delta t \cdot t^5 \cdot \sigma^2. \tag{4.26}$$

Then, the SD of error in position estimation is

$$\sigma_{px}(t) = \frac{\sqrt{5}}{10} f_z^i \cdot \text{ARW} \cdot t^{5/2}. \tag{4.27}$$

Table 4.4 Propagation of position errors in 2D strapdown inertial navigation due to stochastic errors.

Stochastic error source	SD of position estimation error	
	Along x-axis	Along z-axis
ARW	$\frac{\sqrt{5}}{10}f_z^i \cdot \text{ARW} \cdot t^{5/2}$	$\frac{\sqrt{5}}{10}f_x^i \cdot \text{ARW} \cdot t^{5/2}$
VRW	$\frac{\sqrt{3}}{3}\, \text{VRW} \cdot t^{3/2}$	$\frac{\sqrt{3}}{3}\, \text{VRW} \cdot t^{3/2}$
RRW	$\frac{\sqrt{7}}{42}f_z^i \cdot \text{RRW} \cdot t^{7/2}$	$\frac{\sqrt{7}}{42}f_x^i \cdot \text{RRW} \cdot t^{7/2}$
AcRW	$\frac{\sqrt{5}}{10}\, \text{AcRW} \cdot t^{5/2}$	$\frac{\sqrt{5}}{10}\, \text{AcRW} \cdot t^{5/2}$

Similar analysis can be conducted on VRW, RRW, and AcRW, and the results are summarized in Table 4.4. Once again, the values in Table 4.4 only qualitatively show how position error propagates due to different error sources. In practical navigation applications, the trend is the same but values will be different.

4.3.2 Error Propagation in Navigation Frame

The equations for error propagation in a full 3D n-frame can be derived by taking derivative of (3.9) and (3.13), and neglecting the higher-order terms, if we assume that navigation errors are small compared to the actual values. However, an analytical expression for the error of estimation is usually not available due to the unknown and usually complex dynamics of motion during navigation. The results are given without proof below, and more detailed material can be found in [25].

$$\dot{\delta\phi} \approx -\Omega_{in}^n \delta\phi + \delta\omega_{in}^n - C_b^n \delta\omega_{ib}^b, \tag{4.28}$$

$$\dot{\delta v} \approx [f^n \times]\delta\phi + C_b^n \delta f^b - (2\omega_{ie}^n + \omega_{en}^n) \times \delta v - (2\delta\omega_{ie}^n + \delta\omega_{en}^n) \times v - \delta g, \tag{4.29}$$

$$\dot{\delta p} = \delta v, \tag{4.30}$$

where $\delta\phi$ is the error of orientation estimation in roll, pitch, and yaw angles, δv is the error of velocity estimation in the n-frame, and δp is the error of position estimation in the n-frame. The above equations can be written in the state-space form:

$$\frac{d}{dt}\begin{bmatrix} \delta\phi \\ \delta v \\ \delta p \\ b_g^b \\ b_a^b \end{bmatrix} = \begin{bmatrix} -[\omega_{in}^n\times] & F_{\delta v}^{\delta\theta} & 0 & -C_b^n & 0 \\ [f^n\times] & C_1 & C_2 & 0 & C_b^n \\ 0 & I & 0 & 0 & 0 \\ 0 & 0 & 0 & 0 & 0 \\ 0 & 0 & 0 & 0 & 0 \end{bmatrix}\begin{bmatrix} \delta\phi \\ \delta v \\ \delta p \\ b_g^b \\ b_a^b \end{bmatrix} + \begin{bmatrix} C_b^n \cdot \epsilon_{\text{ARW}} \\ C_b^n \cdot \epsilon_{\text{VRW}} \\ 0 \\ \epsilon_{\text{RRW}} \\ \epsilon_{\text{AcRW}} \end{bmatrix}, \tag{4.31}$$

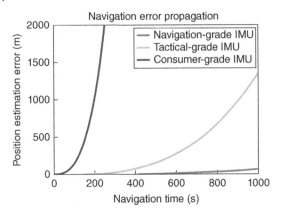

Figure 4.6 Propagation of navigation error with different grades of IMUs.

where b_g^b is the gyroscope bias expressed in b-frame, b_a^b is the accelerometer bias expressed in the b-frame, ϵ_{ARW} is the ARW of the gyroscopes, and ϵ_{VRW} is the VRW of the accelerometers. ϵ_{RRW} and ϵ_{AcRW} are noise terms in the first order Markov process in the modeling of the bias b_g^b and b_a^b, i.e. RRW and AcRW, respectively, and $F_{\delta v}^{\delta \theta}$ is the term related to transport rate, and it is expressed as

$$F_{\delta v}^{\delta \theta} = \frac{1}{R} \begin{bmatrix} 0 & 1 & 0 \\ -1 & 0 & 0 \\ 0 & -\tan L & 0 \end{bmatrix}, \tag{4.32}$$

where R is the Earth radius, L is the latitude of the system, and C_1 and C_2 are the terms related to the Coriolis effects due to both the Earth rotation and the transport rate, and they correspond to the third and fourth term of the right-hand-side of (4.29).

Figure 4.6 shows the simulation result of the propagation of navigation error with different grades of IMUs. The nominal trajectory is a straight line toward the North. For consumer grade IMU (gyroscope BI > 100 °/h, accelerometer BI > 50 mg), the navigation error reaches 10 m within 10 seconds. Even for navigation-grade IMU (gyroscope BI < 0.01 °/h, accelerometer BI < 0.05 mg), error of position estimation reaches 70 m within 1000 seconds, or equivalently about 1 nautical mile per hour, but the speed at which the position estimation error grows increases as time increases. Therefore, we conclude that to achieve the long-term inertial navigation, or to achieve high navigation accuracy with relatively low IMU performances, aiding techniques are needed.

4.4 Conclusions

In this chapter, the navigation error analysis was conducted. The inertial sensor errors and IMU assembly errors were first introduced and modeled. Then, the IMU calibration methods were briefly presented as one of the main methods for navigation error reduction. Next, the navigation error propagation was studied, relating the IMU errors to the navigation errors. Finally, an estimate of navigation error as a function of navigation time was given for different grades of IMU, indicating that for pedestrian inertial navigation where consumer grade or tactical grade IMUs are commonly used, aiding techniques are necessary. Different aiding techniques will be introduced and analyzed in the following chapters.

References

1 El-Sheimy, N., Hou, H., and Niu, X. (2008). Analysis and modeling of inertial sensors using Allan variance. *IEEE Transactions on Instrumentation and Measurement* 57 (1): 140–149.

2 IEEE Std 962-1997 (R2003) (2003). *Standard Specification Format Guide and Test Procedure for Single-Axis Interferometric Fiber Optic Gyros, Annex C.* IEEE.

3 Flenniken, W.S., Wall, J.H., and Bevly, D.M. (2005). Characterization of various IMU error sources and the effect on navigation performance. *ION GNSS*, Long Beach, CA, USA (13–16 September 2005).

4 Bancroft, J.B. and Lachapelle, G. (2012). Estimating MEMS gyroscope g-sensitivity errors in foot mounted navigation. *IEEE Ubiquitous Positioning, Indoor Navigation, and Location Based Service (UPINLBS)*, Helsinki, Finland (3–4 October 2012).

5 Hayal, A.G. (2010). Static Calibration of Tactical Grade Inertial Measurement Units. *Report No. 496.* Columbus, OH: The Ohio State University.

6 Poddar, S., Kumar, V., and Kumar, A. (2017). A comprehensive overview of inertial sensor calibration techniques. *Journal of Dynamic Systems, Measurement, and Control* 139 (1): 011006.

7 Lefevre, H.C. (2014). *The Fiber-Optic Gyroscope*, 2e. Artech House.

8 Passaro, V., Cuccovillo, A., Vaiani, L. et al. (2017). Gyroscope technology and applications: a review in the industrial perspective. *Sensors* 17 (10): 2284.

9 Yazdi, N., Ayazi, F., and Najafi, K. (1998). Micromachined inertial sensors. *Proceedings of the IEEE* 86 (8): 1640–1659.

10 Petovello, M.G., Cannon, M.E., and Lachapelle, G. (2004). Benefits of using a tactical-grade IMU for high-accuracy positioning. *Navigation* 51 (1): 1–12.

11 Bosch (2020). BMI160 Datasheet. https://ae-bst.resource.bosch.com/media/_tech/media/datasheets/BST-BMI160-DS000.pdf (accessed 08 March 2021).

12 StMicroelectronics (2018). ISM330DLC Datasheet. https://www.st.com/resource/en/datasheet/ism330dlc.pdf (accessed 08 March 2021).

13 InvenSense (2020). ICM-42605 Datasheet. http://www.invensense.com/wp-content/uploads/2019/04/DS-ICM-42605v1-2.pdf (accessed 08 March 2021).

14 Analog Devices (2020). ADI16495 Datasheet. https://www.analog.com/media/en/technical-documentation/data-sheets/ADIS16495.pdf (accessed 08 March 2021).

15 Honeywell (2020). HG1930 Datasheet. https://aerospace.honeywell.com/en/~/media/aerospace/files/brochures/n61-1637-000-000-hg1930inertialmeasurementunit-bro.pdf (accessed 08 March 2021).

16 Systron Donner (2020). SDI500 Datasheet. https://www.systron.com/sites/default/files/965755_m_sdi500_brochure_0.pdf (accessed 08 March 2021).

17 Northrop Grumman (2013). LN-200S Datasheet. https://www.northropgrumman.com/Capabilities/LN200sInertial/Documents/LN200S.pdf (accessed 08 March 2021).

18 Safran (2016). PRIMUS 400 Datasheet. https://www.safran-electronics-defense.com/security/navigation-systems (accessed 08 March 2021).

19 Honeywell (2018). HG9900 Datasheet. https://aerospace.honeywell.com/en/~/media/aerospace/files/brochures/n61-1638-000-000-hg9900inertialmeasurementunit-bro.pdf (accessed 08 March 2021).

20 GEM elettronica (2020). IMU-3000 Datasheet. http://www.gemrad.com/imu-3000/ (accessed 08 March 2021).

21 Syed, Z.F., Aggarwal, P., Goodall, C. et al. (2007). A new multi-position calibration method for MEMS inertial navigation systems. *Measurement Science and Technology* 18 (7): 1897–1907.

22 Shin, E.-H. and El-Sheimy, N. (2002). A new calibration method for strapdown inertial navigation systems. *Journal for Geodesy, Geoinformation and Land Management* 127: 1–10.

23 Krakiwsky, E.J. (1990). The Method of Least Squares: A Synthesis of Advances. *Lecture notes, UCGE Reports 10003*. University of Calgary.

24 Woodman, O.J. (2007). An Introduction to Inertial Navigation. *No. UCAM-CL-TR-696*. University of Cambridge, Computer Laboratory.

25 Titterton, D. and Weston, J. (2004). *Strapdown Inertial Navigation Technology*, 2e, vol. 207. AIAA.

5

Zero-Velocity Update Aided Pedestrian Inertial Navigation

As discussed in Chapter 4, the navigation errors accumulate polynomially with respect to time in strapdown inertial navigation, and current Inertial Measurement Unit (IMU) technologies cannot reach the desirable performance to support the pedestrian inertial navigation (see Figure 5.1). Therefore, aiding techniques are needed to constrain the navigation error growth. In this chapter, we focus on the self-contained aiding techniques for pedestrian inertial navigation, which can limit the navigation error propagation of the strapdown inertial navigation while keeping the whole system still independent of the environment.

Compared to general inertial navigation applications where the motion of the system is typically unpredictable, there are observable features in the motion of the system in pedestrian inertial navigation, and these features are typically periodic due to dynamics of the human gait. Therefore, navigation drift compensation can be conducted by taking advantage of the known features or the kinematic relation in human gait cycle. For example, it has been demonstrated that the stride length in human walking can be correlated to the gait frequency [1], the vertical acceleration during walking [2], and opening angles of the thigh [3]. Furthermore, the motion of the foot will periodically return to stationary when it is on the floor, and this information can also be used to restrain the navigation error propagation.

In this chapter, we will start focusing on pedestrian inertial navigation, instead of general inertial navigation applications. One of the most commonly used aiding techniques in the pedestrian inertial navigation is the Zero-Velocity Update (ZUPT) aiding. The concept, algorithmic and hardware implementation, and parameter selection of pedestrian inertial navigation are discussed in this chapter.

5.1 Zero-Velocity Update Overview

In pedestrian inertial navigation, the IMU can be mounted on different parts of body to take advantage of different motion patterns, such as foot, shank, thigh,

Pedestrian Inertial Navigation with Self-Contained Aiding, First Edition. Yusheng Wang and Andrei M. Shkel.
© 2021 The Institute of Electrical and Electronics Engineers, Inc. Published 2021 by John Wiley & Sons, Inc.

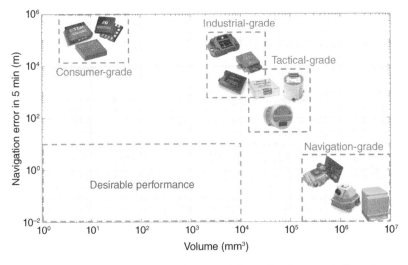

Figure 5.1 Relation between the volumes and the navigation error in five minutes of IMUs of different grades. The dashed box in the lower left corner indicates the desired performance for the pedestrian inertial navigation, showing the need for aiding techniques.

waist, shoulder, and head. Among all the mentioned mounting positions, foot is the most commonly used location due to its simple motion during walking. The foot periodically returns to stationary state when it is on the floor during the stance phase of a walking cycle. The stationary state can be used to limit the long-term velocity and angular rate drift, thus greatly reduce the navigation error. The most commonly used algorithm is the ZUPT aiding algorithm. This algorithm takes advantage of the stationary state of the foot when it is on the floor (stance phase), and uses the zero-velocity information to compensate for navigation errors. In its implementation, IMU is fixed on the foot to perform navigation and detect the stance phase at the same time. Whenever the stance phase is detected, the zero-velocity information of the foot will be input into the system as a pseudo-measurement to compensate for IMU biases and navigation errors, thus reducing the navigation error growth in the system. One of the main advantages of ZUPT is its ability to obtain pseudo-measurement of the velocity, which is otherwise unobservable by IMUs.

The effects of the ZUPT aiding are demonstrated in Figure 5.2. Figure 5.2a shows the estimated velocity of the system along the North. In the strapdown inertial navigation without ZUPT aiding, the estimated velocity drifts away over time due to the IMU errors. However, with the ZUPT implementation, the estimated velocity is close to zero, and as shown in Figure 5.2b, the corresponding estimated trajectory is also closed to the ground truth, which is "8" in shape.

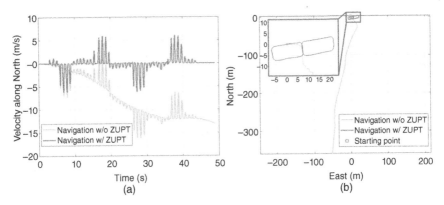

Figure 5.2 Comparison of the estimated velocity of the North and estimated trajectory for navigation with and without ZUPT aiding. Source: Data from OpenShoe [4].

The advantages of the zero-velocity aiding include:

- Only one IMU is used for both inertial navigation and ZUPT aiding, and no extra sensing modality is needed. Therefore, the ZUPT-aided pedestrian inertial navigation can be achieved with simple hardware implementation.
- The algorithm is realized through the Extended Kalman Filter (EKF), which is the optimal linear estimator and requires little computational power. Thus, the algorithm can be achieved by miniaturized system.
- The EKF is able to estimate both the navigation errors and the IMU errors, which can be used to compensate for raw IMU readout data and further improve the final navigation accuracy.

The first few publications demonstrating the concept, implementation, and experimental results of the ZUPT algorithm include [5–7], and more details of the algorithm and implementation were reported in [8]. Many other studies have been conducted to further understand its implementation, characterize its performance, analyze its limitation, relate the navigation error with the IMU performances, and commercialize it [9–12].

There are two key parts involved in the ZUPT-aided navigation algorithm: the stance phase detector and the pseudo-measurement of the motion of the foot.

A stance phase detector is needed to detect the event when the foot is on the floor. Mathematically, the stance phase detector can be formulated as a binary hypothesis testing problem [13]. Generalized Likelihood Ratio Test (GLRT) can be conducted to form the detector. Different hypothesis can be made on the features of the IMU readout during the stance phase, and different stance phase detectors can be developed accordingly. The simplest type is called acceleration moving variance detector, where a stance phase is detected if the variance of the accelerometer readout over a short period of time is small [11]. Equivalently, the change in accelerometer readout can also be directly used as an indicator [8]. Another type of the stance phase detector is called the acceleration magnitude detector, which determines that the IMU is stationary if the magnitude of the measured specific force vector is close to the gravity [14]. Gyroscope readout can also be used in the stance phase detection. Angular rate energy detector detects the stance phase if the root-mean-square of the gyroscope readout is small [15]. A more widely used detector is called Stance Hypothesis Optimal dEtector (SHOE), which is a combination of acceleration magnitude detector and angular rate energy detector, where both accelerometer and gyroscope readouts are utilized [16]. It has been shown that the SHOE detector outperforms marginally the angular rate energy detector, and the acceleration moving variance detector and the acceleration magnitude detector are not as good as the previous two [13].

After identifying the stance phase, pseudo-measurement on the motion of the foot during the stance phase is fed into the EKF, and different types of pseudo-measurement have been reported. The most commonly used pseudo-measurement is the zero-velocity information, where the velocity of the system is simply set to be zero. The simple pseudo-measurement is associated with a linear measurement model and no extra parameter is needed in the system [17]. More complicated models have been proposed to simulate the motion of the foot during the stance phase. For example in [18], a pure rotation of the foot around a pivot point near the toes was assumed during the stance phase. In this model, the rotational speed during the stance phase could be extracted from the gyroscope readout, and the distance between the pivot point and the IMU could be measured or manually tuned. A better navigation accuracy was demonstrated with the rotational pseudo-measurement during the stance phase at the cost of nonlinear measurement model and more parameters involved. Zero-Angular-Rate-Update (ZARU) has also been proposed, where a zero angular rate of the foot is used as a pseudo-measurement during the stance phases [19]. ZARU provides the system more observability of the yaw angle, but it is shown to be accurate enough only when the subject stands still [7].

5.2 Zero-Velocity Update Algorithm

5.2.1 Extended Kalman Filter

Kalman filter is a widely applied estimation tool in time series analysis and is used in fields such as signal processing, navigation, and motion planning. It is the optimal linear estimator for linear system with additive independent white noise [20]. Kalman filter fuses different measurements of the system associated with their measurement uncertainties, and estimates unknown system states that tend to be more accurate than the results obtained from any single measurement. However, most systems are nonlinear in practice, and EKF has been proposed to linearize the nonlinear system about the estimate of the current mean and covariance.

The EKF works in a system with the following given discrete state transition and measurement models

$$
\begin{aligned}
x_k &= f(x_{k-1}, u_k) + v_k, \\
z_k &= h(x_k) + w_k,
\end{aligned}
\tag{5.1}
$$

where k is the time step, $f(\cdot)$ is the state transition function, $h(\cdot)$ is the measurement function, x_k is the system state, u_k is the control input, z_k is the measurement, v_k is the process noise, and w_k is the measurement noise; both are zero-mean Gaussian noises with covariance Q_k and R_k, respectively. Note that $f(\cdot)$ and $h(\cdot)$ may not be linear.

There are two steps in the EKF: the predict step and the update step. In the predict step, the EKF provides a state estimate of the current time step (called the *a priori* state estimate) based on the state estimate from the previous time step and the current control input. In the update step, the EKF combines the *a priori* estimate with the current measurement to adjust the state estimate. The adjusted estimate is called the *a posteriori* state estimate. The update step can be skipped if the measurement is unavailable.

The EKF proceeds as follows:

Predict step:

a priori state estimate $\qquad \hat{x}_{k|k-1} = f(\hat{x}_{k-1|k-1}, u_k) + v_k,$ (5.2)

a priori estimate error covariance $\qquad P_{k|k-1} = F_k P_{k-1|k-1} F_k^{\mathrm{T}} + Q_k.$ (5.3)

Update step:

Measurement residual $\qquad v_k = z_k - h(\hat{x}_{k|k-1}),$ (5.4)

Kalman gain $\qquad W_k = P_{k|k-1} H_k^{\mathrm{T}} (H_k P_{k|k-1} H_k^{\mathrm{T}} + R_k)^{-1},$ (5.5)

a posteriori state estimate $\qquad \hat{x}_{k|k} = \hat{x}_{k|k-1} + W_k v_k,$ (5.6)

a posteriori estimate error covariance $\quad P_{k|k} = (I - W_k H_k) P_{k|k-1},$ (5.7)

where $\hat{x}_{m|n}$ denotes the estimate of x at time step m given measurements up to time step n, and $P_{m|n}$ denotes the corresponding estimate covariance, and

$$
F_k = \frac{\partial f}{\partial x}\Big|_{\hat{x}_{k-1|k-1}, u_k}, \qquad H_k = \frac{\partial h}{\partial x}\Big|_{\hat{x}_{k|k-1}}.
$$

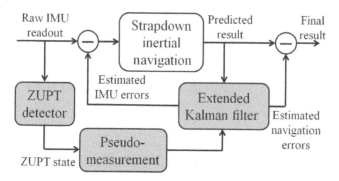

Figure 5.3 Diagram of the ZUPT-aided pedestrian inertial navigation algorithm.

5.2.2 EKF in Pedestrian Inertial Navigation

In the pedestrian inertial navigation, the EKF is commonly used to fuse the IMU readouts with other aiding techniques to obtain a more accurate navigation result. In most cases, instead of the navigation state itself, the EKF is designed to work on the navigation errors as the state space, in order to avoid the issues related to the linearization of the highly nonlinear motion dynamics [21].

The diagram of pedestrian inertial navigation with the EKF is presented in Figure 5.3. The strapdown inertial navigation algorithm is first applied on IMU readouts to predict the current navigation states (position, velocity, and orientation), while in the EKF, the navigation errors are used as the system state. Whenever other aiding techniques (for example, ZUPT, biomechanical modeling, ranging) are available, their readouts are compared with the predicted navigation states, and the difference is used as the measurement in the EKF to update the system state. The EKF is able to estimate both the IMU errors, which can be used as a feedback to compensate for the raw IMU input, and the navigation errors, which is directly used to update the navigation result.

5.2.3 Zero-Velocity Update Implementation

In the ZUPT-aided pedestrian inertial navigation, a standard strapdown Inertial Navigation System (INS) mechanization is implemented in the navigation frame, and the details are introduced in Chapter 3. At the same time, the EKF is performed for error estimation and compensation. In most ZUPT-aided pedestrian inertial navigation algorithm, the error states are used as the system states

$$\delta \vec{x} = [\delta \vec{\theta}^{\mathrm{T}}, \delta \vec{v}_n^{\mathrm{T}}, \delta \vec{s}_n^{\mathrm{T}}, \delta \vec{x}_g^{\mathrm{T}}, \delta \vec{x}_a^{\mathrm{T}}]^{\mathrm{T}}, \tag{5.8}$$

where $\delta\vec{\theta}$ is the three-axis attitude error in the navigation frame, $\delta\vec{v}_n$ and $\delta\vec{s}_n$ are the vectors of velocity and position errors along the North, East, and Down directions of the navigation coordinate frame, $\delta\vec{x}_g$ is the gyroscope states (12-element vector) modeling gyroscope bias, scale factor error, misalignment, and non-orthogonality, and $\delta\vec{x}_a$ is the accelerometer states (9-element vector) modeling accelerometer bias, scale factor error, and non-orthogonality [22]. A full dynamic error model can be approximated by

$$
\delta\dot{\vec{x}} = \begin{bmatrix} -[\vec{\omega}_i^n \times] & F_{\delta v}^{\delta\theta} & 0 & -F_g & 0 \\ [\vec{f}^n \times] & C_1 & C_2 & 0 & F_a \\ 0 & I & 0 & 0 & 0 \\ 0 & 0 & 0 & 0 & 0 \\ 0 & 0 & 0 & 0 & 0 \end{bmatrix} \delta\vec{x} + \begin{bmatrix} C_{s_g}^n \cdot \epsilon_{\mathrm{ARW}} \\ C_{s_a}^n \cdot \epsilon_{\mathrm{VRW}} \\ 0 \\ \epsilon_{b_g} \\ \epsilon_{b_a} \end{bmatrix}, \tag{5.9}
$$

where $[\vec{\omega}_i^n \times]$ and $[\vec{f}^n \times]$ are the skew-symmetric cross-product-operators of angular rate of the navigation frame relative to the inertial frame expressed in the navigation frame, and of accelerometer output in the navigation frame, respectively. I is the identity matrix. $F_{\delta v}^{\delta\theta}$ is the term related to transport rate, C_1 and C_2 are the terms related to the Coriolis effects due to the Earth rotation and the transport rate, $C_{s_g}^n$ and $C_{s_a}^n$ are the Direction Cosine Matrices (DCM) from the navigation frame to the coordinate frames of accelerometers and gyroscopes, respectively. F_g and F_a are matrices (3 by 12) and (3 by 9) modeling the linearized dynamics of the states $\delta\vec{x}_g$ and $\delta\vec{x}_a$, respectively, ϵ_{ARW} is the Angle Random Walk (ARW) of the gyroscopes and ϵ_{VRW} is the Velocity Random Walk (VRW) of the accelerometers. ϵ_{b_g} and ϵ_{b_a} are noise terms in the first order Gauss Markov process in the modeling of the states $\delta\vec{x}_g$ and $\delta\vec{x}_a$, respectively [23].

For a typical IMU, scale factor errors, misalignments, and non-orthogonalities vary slowly during a single pedestrian inertial navigation process and can be approximated by bias errors. We assume that the calibration process before the navigation is able to remove all deterministic biases. Therefore, for pedestrian inertial navigation, only stochastic bias errors and white noises (ARW for gyroscopes and VRW for accelerometers) need to be considered to achieve a high enough accuracy. The Earth rotation and the transport rate are also neglected. The simplification of the error model yields a shorter system state

$$
\delta\vec{x} = [\delta\vec{\theta}^{\mathrm{T}}, \delta\vec{v}_n^{\mathrm{T}}, \delta\vec{s}_n^{\mathrm{T}}, \delta\vec{b}_g^{\mathrm{T}}, \delta\vec{b}_a^{\mathrm{T}}]^{\mathrm{T}}, \tag{5.10}
$$

where $\delta\vec{b}_g$ is the bias of three gyroscopes and $\delta\vec{b}_g$ is the bias of three accelerometers. Note that biases are not systematic errors. With these corrections, the dynamic error model becomes

$$\dot{\delta\vec{x}} = \begin{bmatrix} 0 & 0 & 0 & -C_b^n & 0 \\ [\vec{f}^n \times] & 0 & 0 & 0 & C_b^n \\ 0 & I & 0 & 0 & 0 \\ 0 & 0 & 0 & 0 & 0 \\ 0 & 0 & 0 & 0 & 0 \end{bmatrix} \delta\vec{x} + \begin{bmatrix} C_b^n \cdot \epsilon_{\text{ARW}} \\ C_b^n \cdot \epsilon_{\text{VRW}} \\ 0 \\ \epsilon_{\text{RRW}} \\ \epsilon_{\text{AcRW}} \end{bmatrix} \triangleq A\delta\vec{x} + B, \qquad (5.11)$$

where C_b^n is the DCM from the navigation frame to the body frame, which is assumed to be aligned with the sensor frame, ϵ_{RRW} is Rate Random Walk (RRW) of the gyroscopes, and ϵ_{AcRW} is Accelerometer Random Walk (AcRW) of the accelerometers.

For each time step, the predict step in the EKF is necessary: besides calculating the system states (position, velocity, and attitude) in the standard strapdown navigation algorithm, *a priori* error covariance is propagated according to (5.3), with F and Q defined as

$$F = \exp\{A \cdot \Delta t\} \approx I + A \cdot \Delta t,$$

$$Q = \text{Var}\{BB^{\text{T}}\} \cdot \Delta t,$$

where Δt is the length of each time step, and B is the process noise defined in (5.11). In the discrete form, the system state update can be expressed as

$$\delta\vec{x}_{k+1|k} = F_k \cdot \delta\vec{x}_{k|k}.$$

To activate the update part of the EKF, a zero-velocity detector is needed to detect the stance phase in each gait cycle. Standard SHOE can be applied [13]. In this method, both accelerometer readouts and gyroscope readouts are taken into consideration to detect the stance phase. During the stance phase, the magnitude of the specific force should be equal to the local gravity, and the angular rate of the foot should be zero. However, the direction of the gravity is unknown to the detector, and thus its Maximum Likelihood Estimate (MLE) is used. During a time epoch consisting of W number of measurements, the mean square error of the gyroscope readouts, and the mean square error of the accelerometer readouts subtracted by a vector with the magnitude of the gravity and the direction of the averaged specific force are calculated, weighted by the uncertainty of the measurements, summed up, and compared with the threshold T (dimensionless). If the value is less than the threshold, the stance phase is detected. In the case of zero-velocity detection, the effect of false alarm is much worse than that of a miss-detection. Therefore, a proper combination of T and W is needed to minimize the probability of false detection at the cost of some miss-detection of the zero-velocity events. A typical selection of parameters can be W equals 5 and T equals 3×10^4. The detector can be mathematically expressed as

$$\text{ZUPT} = H\left\{ \frac{1}{W} \sum_{k=1}^{W} \left(\frac{\|\mathbf{y}_k^a - g \cdot \mathbf{y}_n^{\bar{a}}\|^2}{\sigma_a^2} + \frac{\|\mathbf{y}_k^\omega\|^2}{\sigma_\omega^2} \right) - T \right\},$$

where ZUPT is the logical indicator of the detector, $H\{\cdot\}$ is the Heaviside step function, $\boldsymbol{y}_k^a$ and $\boldsymbol{y}_k^\omega$ are accelerometer and gyroscope readouts at time step k, respectively, $\overline{\boldsymbol{y}_n^a}$ is the normalized and averaged accelerometer readout within the measurement window, σ_a and σ_ω are the white noise level of the accelerometers and gyroscopes of the IMU, respectively.

When the stance phase is detected, the ZUPT is applied as the pseudo-measurement, and velocity in the system state is considered as the measurement residual $\boldsymbol{v}_k$ to update the state estimation

$$\vec{v}_k = \begin{bmatrix} 0 & I_{3\times 3} & 0 & 0 & 0 \end{bmatrix} \cdot \delta\vec{x}_k + \vec{w}_k \triangleq H \cdot \delta\vec{x}_k + \vec{w}_k,$$

where H is called the observation matrix, and $\vec{w}_k$ is the measurement noise, which is mainly due to nonzero velocity of the IMU during the stance phase [24]. The covariance of $\vec{w}_k$ is denoted by R_k. In most studies, $\vec{w}_k$ is assumed to be white with constant and isotropic standard deviation r, which is generally set in the range from 0.001 to 0.1 m/s [8, 16, 24, 25]. The value r is also called velocity uncertainty. Therefore, the noise covariance matrix can be expressed as $R_k = r^2 I_{3\times 3}$. Although it is common to tune the parameters used in the EKF, a more proper way is to actually measure the distribution of the velocity of the foot during the stance phases, and set the velocity uncertainty accordingly.

After the EKF receives the measurement information, it updates the system state according to (5.5), (5.6), and (5.7).

5.3 Parameter Selection

As mentioned earlier, although parameter tuning is acceptable in the Kalman filter in order to achieve the best estimation result, it is still preferable to set the parameters according to the actual case, if it is available [26]. In this section, we discuss how to determine the level of process noise and measurement noise in the ZUPT-aided pedestrian inertial navigation algorithm. The determination of the measurement noise is the focus of this section, since the process noise can be easily obtained by characterizing the IMU and extracting data from the Allan deviation. These specifications are also available from the IMU datasheet.

Since no actual measurement is conducted in the ZUPT-aided pedestrian inertial navigation, and only pseudo-measurement of zero-velocity is used, the measurement noise can be interpreted as the velocity uncertainty during the stance phase, i.e. the variance of the distribution of the velocity of the foot during the stance phase [24].

A method to estimate the velocity uncertainty during the stance phase was proposed and demonstrated in [24]. In this method, the velocity uncertainty can be estimated based on a foot-mounted IMU, and no other sensing modalities, such as global positioning system (GPS), motion tracker, and velometer, are needed, which greatly facilitates the *in situ* measurement of the velocity uncertainty. There are three main steps in this method: ZUPT-aided inertial navigation algorithm is first implemented to the IMU data to estimate the trajectory and to extract the stance phases. Next, a free strapdown navigation algorithm is applied to the IMU readouts during the stance phases, assuming a zero initial velocity. Finally, distribution of the final velocity is analyzed and its standard deviation is extracted as the velocity uncertainty. The initial orientation of the IMU during the stance phases used in the second step is obtained from previous navigation results in the first step. An example of the velocity propagations during the stance phases is shown in Figure 5.4. The average length of the stance phase in this case is around 0.48 seconds. No obvious features can be extracted along the horizontal directions, indicating a random nature of the velocity during the stance phase. The standard deviation of the final velocity is calculated to be around 0.017 m/s for all three directions, as shown in Figure 5.5. The final velocity distribution is of a bell shape, indicating that it is reasonable to model the velocity uncertainty to be normally distributed. The velocity bias along the vertical direction may be explained as follows. When the zero-velocity detector determines the start of the stance phase, the foot is not fully stationary and the residual vertical velocity is downward, but the initial velocity of the stance phase is assumed to be zero, and the velocity bias is thereby introduced. Even though the velocity uncertainty may not be exactly normally distributed, EKF estimation is still the optimal linear

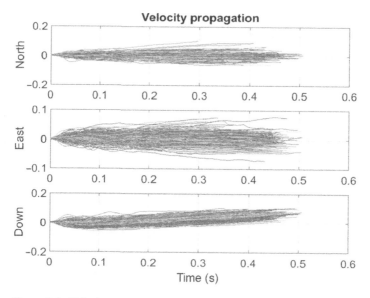

Figure 5.4 Velocity propagation along three orthogonal directions during the 600 stance phases.

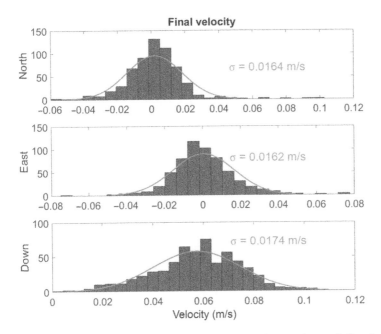

Figure 5.5 Distribution of the final velocity along three orthogonal directions during 600 stance phases. Standard deviation is extracted as the average velocity uncertainty during the stance phase.

estimator with the expectation of the noise to be zero (at least for two horizontal directions).

In order to trace the origin of the velocity uncertainty in the experiment, the velocity uncertainty caused by the white noise of accelerometers and gyroscopes is calculated, since white noise is the dominant IMU error source for short-term navigation.

$$\Delta v_{accel} = \text{VRW} \cdot \sqrt{t} \approx 1 \times 10^{-3} \text{ m/s}, \tag{5.12}$$

$$\Delta v_{gyro} = \sqrt{\frac{1}{3}\text{ARW} \cdot g \cdot t^{3/2}} \approx 8 \times 10^{-5} \text{ m/s}, \tag{5.13}$$

where Δv_{accel} is the velocity uncertainty caused by accelerometer's white noise, Δv_{gyro} is the velocity uncertainty caused by gyroscope's white noise, t is the length of the stance phase, and g is the gravity. According to (5.12) and (5.13), the extracted velocity uncertainty during the stance phase is not related to the IMU noises, since the noise is orders of magnitude lower than what is needed to introduce velocity uncertainty on the order of 0.01 m/s. Therefore, it can be concluded that the velocity uncertainty is caused by the motion and shocks of the foot even though it is on the floor during the stance phase.

5.4 Conclusions

Zero-velocity aiding is one of the most popular aiding techniques used in pedestrian inertial navigation due to its easy implementation and good effect in reducing the navigation error. In this chapter, the concept, algorithmic implementation, and parameter selection of the ZUPT-aided pedestrian inertial navigation were introduced. This chapter is used as a foundation for the following analysis.

References

1 Ladetto, Q. (2000). On foot navigation: continuous step calibration using both complementary recursive prediction and adaptive Kalman filtering. *International Technical Meeting of the Satellite Division of The Institute of Navigation (ION GPS 2000)*, Salt Lake City, UT, USA (19–22 September 2000).

2 Kim, J.W., Jang, H.J., Hwang, D.-H., and Park, C. (2004). A step, stride and heading determination for the pedestrian navigation system. *Journal of Global Positioning Systems* 3 (1–2): 273–279.

3 Diaz, E.M. and Gonzalez, A.L.M. (2014). Step detector and step length estimator for an inertial pocket navigation system. *IEEE International Conference on Indoor Positioning and Indoor Navigation (IPIN)*, Busan, South Korea (27–30 October 2014).

4 OpenShoe (2012). Matlab Implementation. http://www.openshoe.org/?page_id=362 (accessed 08 March 2021).

5 Sher, L. (1996). *Personal Inertial Navigation System (PINS)*. DARPA.

6 Hutchings, L.J. (1998). System and method for measuring movement of objects. US Patent No. 5,724,265.

7 Elwell, J. (1999). Inertial navigation for the urban warrior. *AEROSENSE '99*, Orlando, FL, USA (5–9 April 1999), pp. 196–204.

8 Foxlin, E. (2005). Pedestrian tracking with shoe-mounted inertial sensors. *IEEE Computer Graphics and Applications* 25 (6): 38–46.

9 Ojeda, L. and Borenstein, J. (2007). Personal dead-reckoning system for GPS-denied environments. *IEEE International Workshop on Safety, Security and Rescue Robotics (SSRR)*, Rome, Italy (27–29 September 2007).

10 Nilsson, J.O., Gupta, A.K., and Handel, P. (2014). Foot-mounted inertial navigation made easy. *IEEE International Conference on Indoor Positioning and Indoor Navigation (IPIN)*, Busan, Korea (27–30 October 2014).

11 Godha, S. and Lachapelle, G. (2008). Foot mounted inertial system for pedestrian navigation. *Measurement Science and Technology* 19 (7): 075202.

12 Wang, Y., Chernyshoff, A., and Shkel, A.M. (2019). Study on estimation errors in ZUPT-aided pedestrian inertial navigation due to IMU noises. *IEEE Transactions on Aerospace and Electronic Systems* 56 (3): 2280–2291.

13 Skog, I., Händel, P., Nilsson, J.O., and Rantakokko, J. (2010). Zero-velocity detection — an algorithm evaluation. *IEEE Transactions on Biomedical Engineering* 57 (11): 2657–2666.

14 Krach, B. and Robertson, P. (2008). Integration of foot-mounted inertial sensors into a Bayesian location estimation framework. *IEEE Workshop on Positioning, Navigation and Communication*, Hannover, Germany (27 March 2008), pp. 55–61.

15 Feliz, R., Zalama, E., and Garcia-Bermejo, J.G. (2009). Pedestrian tracking using inertial sensors. *Journal of Physical Agents* 3 (1): 35–43.

16 Skog, I., Nilsson, J.O., and Händel, P. (2010). Evaluation of zero-velocity detectors for foot-mounted inertial navigation systems. *IEEE International Conference on In Indoor Positioning and Indoor Navigation (IPIN)*, Zurich, Switzerland (15–17 September 2010).

17 Ramanandan, A., Chen, A., and Farrell, J.A. (2012). Inertial navigation aiding by stationary updates. *IEEE Transactions on Intelligent Transportation Systems* 13 (1): 235–248.

18 Laverne, M., George, M., Lord, D. et al. (2011). Experimental validation of foot to foot range measurements in pedestrian tracking. *ION GNSS Conference*, Portland, OR, USA (19–23 September 2011).

19 Rajagopal, S. (2008). Personal dead reckoning system with shoe mounted inertial sensors. Master dissertation. KTH Royal Institute of Technology.

20 Kalman, R.E. (1960). A new approach to linear filtering and prediction problems. *Journal of Basic Engineering* 82: 35–45.

21 Hou, X., Yang, Y., Li, F., and Jing, Z. (2011). Kalman filter based on error state variables in SINS + GPS navigation application. *IEEE International Conference on Information Science and Technology*, Nanjing, China (26–28 March 2011), pp. 770–773.

22 Savage, P.G. (2007). *Strapdown Analytics*, 2e. Maple Plain, MN: Strapdown Associates.

23 Huddle, J.K. (1978). Theory and performance for position and gravity survey with an inertial system. *Journal of Guidance, Control, and Dynamics* 1 (3): 183–188.

24 Wang, Y., Askari, S., and Shkel, A.M. (2019). Study on mounting position of IMU for better accuracy of ZUPT-aided pedestrian inertial navigation *IEEE International Symposium on Inertial Sensors and Systems (Inertial)*, Naples, FL, USA (1–5 April 2019).

25 Ren, M., Pan, K., Liu, Y. et al. (2016). A novel pedestrian navigation algorithm for a foot-mounted inertial-sensor-based system. *Sensors* 16 (1): 139.

26 Matisko, P. and Havlena, V. (2010). Noise covariances estimation for Kalman filter tuning. *IFAC Proceedings* 43 (10): 31–36.

6

Navigation Error Analysis in the ZUPT-Aided Pedestrian Inertial Navigation

In Chapter 5, the Zero-Velocity Update (ZUPT)-aided pedestrian inertial navigation has been introduced and demonstrated to be able to greatly reduce the navigation error. However, due to the self-contained nature of inertial navigation, the navigation error still propagates over time. In order to estimate the navigation error growth and to identify the dominant error source, an analytical expression of navigation error propagation in the ZUPT-aided pedestrian inertial navigation is desirable. However, due to the zero-velocity pseudo-measurement and the corresponding error compensation, the overall error propagation will be more complicated than in the case of strapdown inertial navigation without aiding, which is presented in Chapter 4.

In this chapter, navigation errors in ZUPT-aided pedestrian inertial navigation due to Inertial Measurement Unit (IMU) noises are analyzed in detail. It is envisioned that this framework will be a tool to aid in analysis of the effect of errors in sensors, which might lead to a well-informed selection of sensors for the task of ZUPT-aided pedestrian inertial navigation. In this chapter, a 2D biomechanical model to simulate human gait is presented to better understand human walking dynamics and also to serve as the basis for the following numerical simulations.

6.1 Human Gait Biomechanical Model

In this section, we first develop an approach for generation of the foot trajectory. Such biomechanical models are necessary for analytical prediction of errors in the ZUPT-aided pedestrian navigation.

Human ambulatory gait models are multidimensional due to the complex kinematic and dynamic relations between many parts of human body involved during walking. In this section, our focus is only on the trajectories of two feet instead of the whole-body motion. Therefore, a few assumptions are used to simplify the human gait model:

Pedestrian Inertial Navigation with Self-Contained Aiding, First Edition. Yusheng Wang and Andrei M. Shkel.
© 2021 The Institute of Electrical and Electronics Engineers, Inc. Published 2021 by John Wiley & Sons, Inc.

1. The motion of each leg is two-dimensional and parallel to each other, indicating no rotation occurs at the pelvis and no horizontal rotation occurs at the ankles.
2. The dimensions of both legs are identical.
3. The pattern and duration of each step are identical.
4. The floor is flat, resulting in no accumulation of altitude changes during walking.
5. The trajectory is straight; no turning or stopping happens during the navigation.

In the following parts of this section, we first extract the foot motion from the joints rotation in the torso coordinate frame. Next, based on the human gait analysis, the expression of the foot motion is transferred from the torso frame to the navigation frame. Finally, a parameterization is applied to generate a new trajectory with higher order of continuity, while preserving all the key characteristics of the foot motion.

6.1.1 Foot Motion in Torso Frame

The torso frame is a coordinate frame that is fixed to the body trunk. In the torso frame, only the relative motion with respect to the trunk is studied.

Joint movement has been widely studied for pathological purposes, and the angle data are typically extracted by a high-speed camera or wearable sensors [1]. A pattern of joint angle changes was reported in [2] and is reproduced in Figure 6.1a. A simplified human leg model is shown in Figure 6.1b. The leg is modeled as two bars with femur length of 50 cm and tibia length of 45 cm. The foot is modeled as a triangle with the side lengths of 4, 13, and 16 cm, respectively. The parameters are determined by a typical male subject with a height of 180 cm. Position of the forefoot in the torso frame is expressed as

$$x_{\text{forefoot}} = L_1 \sin \alpha + L_2 \sin(\alpha - \beta) + L_3 \sin(\alpha - \beta + \gamma), \tag{6.1}$$

$$y_{\text{forefoot}} = L_1 \cos \alpha + L_2 \cos(\alpha - \beta) + L_3 \cos(\alpha - \beta + \gamma). \tag{6.2}$$

The corresponding parameters are shown in Figure 6.1b. The position of another foot can be calculated by shifting the time by half of a cycle since we assume that every step is identical.

6.1.2 Foot Motion in Navigation Frame

The navigation frame is the coordinate frame that is fixed on the ground with axes pointing to the North, East, and Down directions, respectively. In this frame, the motion of foot with respect to the ground is first studied.

To transfer a foot motion from the torso frame to the navigation frame, the gait analysis is necessary to establish stationary points as a reference in different phases

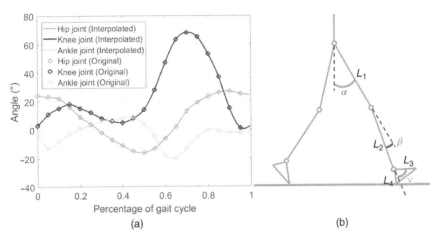

Figure 6.1 (a) Interpolation of joint movement data and (b) simplified human leg model. Source: (a) Modified from Murray et al. [2].

of the gait cycle. Each gait cycle is divided in two phases: stance and swing. The stance phase is a period during which the foot is on the ground. The swing phase is a period when the foot is in the air for the limb advancement [3].

We assume that each gait cycle begins when the left heel contacts the ground (heel strike). During the first 15% of the gait cycle, the left heel is assumed to be stationary and the foot rotates around it (heel rocker) until the whole foot touches the ground. During 15% to 40% of the gait cycle, the whole left foot is on the ground and stationary, and the left ankle joint rotates for limb advancement (ankle rocker). This is also the time period when ZUPT is applied as pseudo-measurements to the Extended Kalman Filter (EKF). For 40% to 60% of the gait cycle, the left heel begins to rise, and this stage ends when the left foot is off the ground. In this stage, the left foot rotates with respect to the forefoot, which we assume to be stationary (forefoot rocker) [4]. The following part of the gait cycle is symmetric to the previous part since we assume that every step is identical. Phases of the gait cycle are presented in Figure 6.2.

After establishing different stationary points in different phases of the gait cycle, we can extract position of the body trunk with respect to the ground. The foot motion can be superimposed on top of the torso motion to obtain the foot motion in the navigation frame.

6.1.3 Parameterization of Trajectory

Abrupt changes of the reference point from the heel to the ankle and to the forefoot will create discontinuity in the trajectory, especially in terms of velocity and

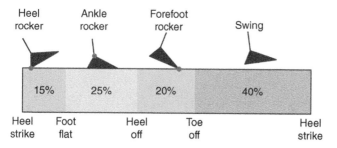

Figure 6.2 Human ambulatory gait analysis. The light gray dots are the stationary points in different phases of one gait cycle.

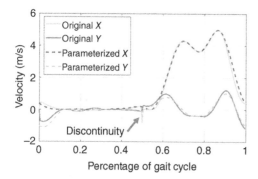

Figure 6.3 Velocity of the parameterized trajectory. A close match is demonstrated and discontinuities were eliminated.

acceleration, as depicted in Figure 6.3. The discontinuities of acceleration result in discontinuous accelerometer readouts, which will cause numerical problems in the algorithm. Therefore, parameterization is needed to generate a new trajectory with a higher order of continuity [5].

The new trajectory to be generated does not have to strictly follow the angle data for each joint and the linkage relations, but ambulatory characteristics should be preserved, especially zero velocity and angular rate during time period of the ankle rocker, in order to conduct the ZUPT-aided pedestrian inertial navigation algorithm on the generated foot trajectory.

The velocity along the trajectory is parameterized to guarantee the continuity of both displacement and acceleration. Key points are first selected to characterize the IMU velocity along the horizontal and vertical directions. For parameterization along the vertical direction, the integral of velocity for a gait cycle should be zero to make sure the altitude does not change over one gait cycle. This is achieved by adjusting velocity values at some of the key points.

The parameterization results are shown in Figure 6.3. The generated velocities (dashed lines) closely follow the original values without losing any characteristics

Figure 6.4 Displacement of the parameterized trajectory. A close match is demonstrated for displacement along the x direction (horizontal). Difference between the displacements along y direction (vertical) is to guarantee the displacement continuity in between the gait cycles.

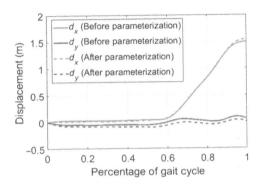

and also eliminate the discontinuity that otherwise would happen in the middle (50%) of the gait cycle, corresponding to a shift of the reference point from the left forefoot to the right heel. The trajectory in terms of position is obtained by integrating the velocity, and the results are shown in Figure 6.4. A close match is demonstrated for displacement along the horizontal direction. For displacement along the vertical direction, the difference is purposefully introduced to guarantee that the altitude of foot does not change after one gait cycle.

6.2 Navigation Error Analysis

The navigation errors in the ZUPT-aided navigation algorithm come mainly from two major sources: systematic modeling errors and IMU noises. We emphasize that in this section, we only analyze the navigation errors caused by IMU noises. In this section, we quantitatively analyze the navigation errors in terms of the estimated orientation, velocity, and position.

6.2.1 Starting Point

A typical propagation of the error in attitude estimation in ZUPT-aided pedestrian inertial navigation and its covariance are presented in Figure 6.5. A similar phenomenon can be observed for the velocity error propagation as well. A few conclusions can be drawn from propagation of errors in attitude estimation:

1. Although the propagated error is random due to the stochastic nature of noise (solid lines in Figure 6.5), the error covariances (bounds) follow a pattern (dashed lines in Figure 6.5);
2. For roll and pitch angles, the covariance reaches a constant and the same level with some fluctuations, but covariance of azimuth angle propagated as $t^{1.5}$ due to Rate Random Walk (RRW) and lack of observability in azimuth angle [6];

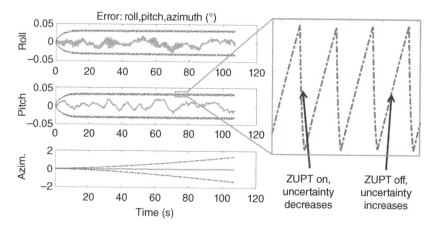

Figure 6.5 A typical propagation of errors in attitude estimations in ZUPT-aided pedestrian inertial navigation. The solid lines are the actual estimation errors, and the dashed lines are the 3σ uncertainty of estimation. Azimuth angle (heading) is the only important EKF state that is not observable from zero-velocity measurements. Source: Wang et al. [7].

3. The covariances are reduced if the update step of the EKF is activated during the stance phase, and they are increased in the predict step during the swing phase.

A starting point of the analysis is an observation that covariances of attitude and velocity reach a stable level with some fluctuation in the long run of the ZUPT-aided navigation algorithm [7], as shown by the dashed lines in Figure 6.5. This observation indicates that in a whole gait cycle, the amount of the covariance increase in the predict step is equal to the amount of the covariance decrease in the update step. Following this observation, we combine the parameters related to ZUPT and the IMU parameters to estimate the overall navigation solution uncertainty. This combination enables us to fully analyze the system behavior and extract the covariance of the error in the system state estimation. For simplicity of derivation, we assume in the analysis a straight-line trajectory toward the North. We also assume a 2D motion of the foot: the foot only moves along the North and Down directions, and the roll and azimuth angles are zero. In case of any other trajectory shapes, the analytical expression for the navigation error may be different, but the general conclusions still hold.

6.2.2 Covariance Increase During Swing Phase

Covariance increases during the swing phase due to the noise in IMU readouts. The *a priori* covariance propagates according to (5.3). To differentiate the error terms in position, velocity, and orientation, we expand (5.3) into 3×3 sub-blocks, suppress subscripts indicating the time steps for simplicity, and use new subscripts to indicate the index of sub-blocks. In this way, subscript 1 corresponds to the angle error, and subscript 2 corresponds to the velocity error, and subscript 3 corresponds to the position error. The *a priori* covariance propagation of angle in a single time step can be estimated by

$$P_{11}^{\text{priori}} \approx P_{11} + Q_{11} - (C_b^n P_{41} + P_{14} C_b^{n^{\text{T}}}) \cdot \Delta t + C_b^n P_{44} C_b^{n^{\text{T}}} \cdot \Delta t^2, \tag{6.3}$$

where Δt is the length of a time step. $C_b^n P_{41}$ and $P_{14} C_b^{n^{\text{T}}}$ are symmetric with respect to each other and share the same on-diagonal terms. Note that in Sections 6.2.2 and 6.2.3, terms in the form of P_{mn} and $P_{mn}(j, k)$ stand for the posterior covariance obtained from previous update step, or *a priori* covariance from previous prediction step if there is no update in the previous step. The last term on the right-hand side of (6.3) can be neglected since the sampling rate is high (typically above 100 Hz). The high sampling frequency also helps to mitigate the error caused by applying the EKF to nonlinear problems. Due to the assumption that the foot motion is 2D, the Direction Cosine Matrix (DCM) can be expressed by

$$C_b^n = \begin{bmatrix} \cos \theta & 0 & \sin \theta \\ 0 & 1 & 0 \\ -\sin \theta & 0 & \cos \theta \end{bmatrix},$$

where θ is the pitch angle of the foot. Since the orientation covariance propagations of the two horizontal directions (roll and pitch) are identical, we only need to focus on one of them. In this study, we select $P_{11}(1, 1)$, which corresponds to the roll angle, and its *a priori* covariance increase during the whole gait cycle is

$$P_{11}^{\text{priori}}(1, 1) \approx P_{11}(1, 1) + (\text{ARW}^2 - 2a_c P_{41}(1, 1)) \cdot t_{\text{stride}}, \tag{6.4}$$

where t_{stride} is the time duration of a gait cycle, a_c is the averaged value of $\cos \theta$ during the whole gait cycle and it is estimated to be around 0.84 for a normal human gait pattern [2]. $P_{41}(3, 1)$ is neglected because it is much smaller than $P_{41}(1, 1)$.

Covariance propagation of the velocity estimation error can be analyzed in a similar way based on (5.3)

$$P_{22}^{\text{priori}} \approx P_{22} + Q_{22} + \{[\vec{f}^n \times]P_{12} + P_{21}[\vec{f}^n \times]^{\text{T}} + C_b^n P_{52} + P_{25} C_b^{n^{\text{T}}}\}\Delta t, \tag{6.5}$$

where P_{12} and P_{21} are symmetric with respect to each other. Integration of (6.5) over a whole gait cycle leads to the increase of the covariance of velocity estimation during a single gait cycle. $[\vec{f}^n \times]$ is composed of two parts: the constant acceleration $\vec{g}$ and the fast-varying acceleration $\vec{a}_m$ caused by motion. The latter can be neglected in the integration because P_{12} is a relatively slowly-varying term compared to $\vec{a}_m$, and therefore, P_{12} can be considered constant in the integral of their multiplication in (6.5) and taken out of the integral. Therefore, the expression becomes an integral of the acceleration $\vec{a}_m$, which equals zero since the velocity returns to its original value in a complete gait cycle. The terms $P_{52}(1,1)$ and $P_{52}(1,3)$ are much smaller than $P_{12}(1,2)$ and thus can be neglected. Therefore, a total *a priori* covariance increase of the velocity error along the North can be expressed as

$$P_{22}^{\text{priori}}(1,1) \approx P_{22}(1,1) + (\text{VRW}^2 - 2g \cdot P_{12}(1,2)) \cdot t_{\text{stride}}, \tag{6.6}$$

where $P_{12}(1,2)$ is the covariance between rotation along the North and the velocity along the East, and g is the gravitational acceleration. $P_{12}(1,2)$ is an important parameter because it corresponds to coupling between the angular rate error and the velocity error. Schuler pendulum, for example, is one of the effects of this term [8]. To complete the analysis, we also need to calculate the covariance increase of $P_{12}(1,2)$. The covariance propagation is described as

$$P_{12}^{\text{priori}}(1,2) \approx P_{12}(1,2) - g \cdot P_{11}(1,1) \cdot t_{\text{stride}}. \tag{6.7}$$

The sub-block in the covariance matrix that corresponds to the position estimation error is P_{33}, and its propagation in the predict step can be expressed as

$$P_{33}^{\text{priori}} = P_{33} + (P_{23} + P_{32}) \cdot \Delta t + P_{22} \cdot \Delta t^2. \tag{6.8}$$

The position estimation uncertainties along the North and the East are represented by $P_{33}(1,1)$ and $P_{33}(2,2)$, respectively, and they only depend on the propagation of $P_{23}(1,1)$ and $P_{23}(2,2)$, which correspond to the coupling between the velocity errors and position errors. Propagations of the coupling terms are expressed as

$$P_{23}^{\text{priori}}(1,1) \approx P_{23}(1,1) + [P_{22}(1,1) + (g - a_D)P_{13}(2,1)] \cdot \Delta t, \tag{6.9}$$

$$P_{23}^{\text{priori}}(2,2) \approx P_{23}(2,2) + [P_{22}(2,2) + (g - a_D)P_{13}(2,1) - a_N P_{13}(3,2)] \cdot \Delta t, \tag{6.10}$$

where a_N is the acceleration along the North, and a_D is the acceleration toward the Down direction. Note that terms related to P_{53} are neglected and not shown in these equations. The reason why a_N and a_D cannot be neglected as in (6.6) will be explained later in this section. The only difference between the two directions is the last term in (6.10).

Similarly, propagations of covariance of other terms are

$$P_{12}^{\text{priori}}(3,2) = P_{12}(3,2) - P_{11}(3,3) \cdot a_N \cdot \Delta t, \tag{6.11}$$

$$P_{13}^{\text{priori}}(2,1) = P_{13}(2,1) + P_{12}(2,1) \cdot \Delta t, \tag{6.12}$$

$$P_{13}^{\text{priori}}(3,2) = P_{13}(3,2) + [P_{12}(3,2) + \sin\theta P_{43}(1,2) - \cos\theta P_{43}(3,2)] \cdot \Delta t. \tag{6.13}$$

$$P_{41}^{\text{priori}}(1,1) = P_{41}(1,1) - P_{44}(1,1)\cos\theta \cdot \Delta t, \tag{6.14}$$

$$P_{41}^{\text{priori}}(1,3) = P_{41}(1,3) + P_{44}(1,1)\sin\theta \cdot \Delta t, \tag{6.15}$$

$$P_{41}^{\text{priori}}(3,3) = P_{41}(3,3) - P_{44}(3,3)\cos\theta \cdot \Delta t, \tag{6.16}$$

$$P_{42}^{\text{priori}}(1,2) = P_{42}(1,2) + P_{41}(1,1) \cdot (-g + a_D) \cdot \Delta t, \tag{6.17}$$

$$P_{42}^{\text{priori}}(3,2) = P_{42}(3,2) - P_{41}(3,3) \cdot a_N \cdot \Delta t, \tag{6.18}$$

$$P_{43}^{\text{priori}}(1,2) = P_{43}(1,2) + P_{42}(1,2) \cdot \Delta t, \tag{6.19}$$

$$P_{43}^{\text{priori}}(3,2) = P_{43}(3,2) + P_{42}(3,2) \cdot a_N \cdot \Delta t, \tag{6.20}$$

$$P_{44}^{\text{priori}}(2,2) = P_{44}(2,2) + \text{RRW}^2 \cdot \Delta t. \tag{6.21}$$

6.2.3 Covariance Decrease During the Stance Phase

During the stance phase, the ZUPT-aided navigation algorithm compensates for the IMU errors/noise, and therefore reduces the covariance of the state estimation. The amount of the total reduction can be calculated based on (5.5) and (5.7).

We first analyze the covariance of the angle estimation. For each time step during the stance phase, the *a posteriori* covariance change can be expressed as

$$
\begin{aligned}
P_{11}^{\text{posteriori}}(1,1) = P_{11}(1,1) &- \frac{P_{12}(1,1)P_{21}(1,1)}{P_{22}(1,1)+r^2} - \frac{P_{12}(1,2)P_{21}(2,1)}{P_{22}(2,2)+r^2} \\
&- \frac{P_{12}(1,3)P_{21}(3,1)}{P_{22}(3,3)+r^2} \\
\approx P_{11}(1,1) &- \frac{P_{12}(1,2)^2}{r^2}.
\end{aligned} \tag{6.22}
$$

In the strapdown inertial navigation mechanization, the rotation along the North is strongly coupled with the acceleration along the East due to the gravity. Therefore, $P_{12}(1,1)$ and $P_{12}(1,3)$ are much smaller than $P_{12}(1,2)$ and can be neglected. The velocity measurement uncertainty is generally much greater than velocity error induced by IMU noises in ZUPT-aided navigation process [5, 9]. As a result, P_{22} in the denominator is much smaller than r^2 and can be neglected.

Similarly, *a posteriori* covariance of other terms that are needed in the derivation can be calculated as

$$P_{12}^{\text{posteriori}}(1,2) = P_{12}(1,2) - P_{12}(1,2) \cdot P_{22}(2,2)/r^2, \tag{6.23}$$

$$P_{22}^{\text{posteriori}}(2,2) = P_{22}(2,2) - P_{22}(2,2)^2/r^2, \tag{6.24}$$

$$P_{13}^{\text{posteriori}}(3,2) = P_{13}(3,2) - P_{23}(2,2) \cdot P_{12}(3,2)/r^2, \tag{6.25}$$

$$P_{13}^{\text{posteriori}}(2,1) = P_{13}(2,1) - P_{23}(1,1) \cdot P_{12}(2,1)/r^2, \tag{6.26}$$

$$P_{33}^{\text{posteriori}}(1,1) = P_{33}(1,1) - P_{23}(1,1)^2/r^2, \tag{6.27}$$

$$P_{22}^{\text{posteriori}}(1,1) = P_{22}(1,1) - P_{22}(1,1)^2/r^2, \tag{6.28}$$

$$P_{33}^{\text{posteriori}}(2,2) = P_{33}(2,2) - P_{23}(2,2)^2/r^2, \tag{6.29}$$

$$P_{44}^{\text{posteriori}}(2,2) = P_{44}(2,2) - P_{42}(2,1)^2/r^2, \tag{6.30}$$

$$P_{42}^{\text{posteriori}}(1,2) = P_{42}(1,2) - P_{42}(1,2)P_{22}(2,2)/r^2, \tag{6.31}$$

$$P_{41}^{\text{posteriori}}(1,1) = P_{41}(1,1) - P_{42}(1,2)P_{21}(2,1)/r^2. \tag{6.32}$$

6.2.4 Covariance Level Estimation

Since ZUPT-aided navigation algorithm has limited observability of azimuth angle as it is shown in Figure 6.5, the propagation of error in the azimuth angle and in the z-axis gyroscope bias is the same as in strapdown inertial navigation

$$P_{44}(3,3) = \text{RRW}^2 \cdot t, \tag{6.33}$$

$$P_{11}(3,3) = \text{ARW}^2 \cdot t + \frac{\text{RRW}^2}{3} \cdot t^3, \tag{6.34}$$

where t is the total navigation time. In this section, terms in the form of P_{mn} and $P_{mn}(j,k)$ stand for the predicted continuous covariance bounds. Besides, in this section, we only focus on the level of covariance in the long term, and neglect the variance change within a single gait cycle. Therefore, we do not distinguish *a priori* and *a posteriori* covariances anymore.

A combination of (6.21) and (6.30) gives

$$\text{RRW}^2 \cdot t_{\text{stride}} = \frac{P_{42}(2,1)^2}{r^2} \cdot N_{\text{stance}}. \tag{6.35}$$

Since $N_{\text{stance}} = f_s \cdot t_{\text{stance}}$, where f_s is the sampling frequency of the IMU, P_{42} can be expressed as

$$P_{42}(1,2) = -\text{RRW} \left[\frac{r^2 \cdot t_{\text{stride}}}{f_s \cdot t_{\text{stance}}} \right]^{\frac{1}{2}}. \tag{6.36}$$

The minus sign in the equation is due to the fact that a positive gyroscope bias along the East will cause a negative velocity estimation error along the North.

Similarly, a combination of (6.17) and (6.31) gives

$$-P_{44}(1,1) \cdot g \cdot t_{\text{stride}} = \frac{P_{42}(1,2)P_{22}(2,2)}{r^2} f_s \cdot t_{\text{stance}}. \tag{6.37}$$

A combination of (6.4) and (6.22) gives

$$[\text{ARW}^2 - 2a_c P_{41}(1,1)] \cdot t_{\text{stride}} = \frac{P_{12}(1,2)^2}{r^2} f_s \cdot t_{\text{stance}}. \tag{6.38}$$

A combination of (6.6) and (6.28) gives

$$[\text{VRW}^2 - 2g \cdot P_{12}(1,2)] \cdot t_{\text{stride}} = \frac{P_{22}(1,1)^2}{r^2} f_s \cdot t_{\text{stance}}. \tag{6.39}$$

From (6.36)–(6.39), we are able to calculate $P_{22}(1,1)$, which is the root of a quartic equation

$$ax^4 + bx^2 + cx + d = 0 \tag{6.40}$$

with coefficients expressed as

$$a = \left[\frac{f_s \cdot t_{\text{stance}}}{2gr^2 \cdot t_{\text{stride}}} \right]^2, b = -\frac{f_s \cdot t_{\text{stance}} \text{VRW}^2}{2g^2 r^2 \cdot t_{\text{stride}}},$$

$$c = -2a_c \frac{\text{RRW}}{g} \sqrt{\frac{r^2 t_{\text{stride}}}{f_s t_{\text{stance}}}}, d = \frac{\text{VRW}^4}{4g^2} - \frac{\text{ARW}^2 r^2 t_{\text{stride}}}{f_s t_{\text{stance}}}.$$

An analytical solution to (6.40) exists, but it is too complicated and not instructive to present here. Therefore, instead of searching for the analytical expression, we calculate the solution numerically. Note, $P_{22}(1,1)$ is the term in the covariance matrix that corresponds to uncertainty of the velocity estimation along the East. The velocity uncertainty is simply $\sigma_v = \sqrt{P_{22}(2,2)}$. From the equations above, $P_{12}(1,2)$ and $P_{41}(1,1)$ can be calculated as

$$P_{12}(1,2) = -\left[\text{ARW}^2 \frac{r^2 t_{\text{stride}}}{f_s t_{\text{stance}}} + 2a_c \frac{\text{RRW} \cdot \sigma_v^2}{g} \sqrt{\frac{r^2 t_{\text{stride}}}{f_s t_{\text{stance}}}} \right]^{\frac{1}{2}}, \tag{6.41}$$

$$P_{41}(1,1) = -\frac{\text{RRW} \cdot \sigma_v^2}{g} \sqrt{\frac{f_s t_{\text{stance}}}{r^2 t_{\text{stride}}}}. \tag{6.42}$$

A combination of (6.14) and (6.32) gives

$$P_{44}(1,1)a_c \cdot t_{\text{stride}} = \frac{P_{42}(1,2)P_{21}(2,1)}{r^2} f_s \cdot t_{\text{stance}}, \tag{6.43}$$

or equivalently

$$P_{44}(1,1) = \frac{P_{42}(1,2)P_{21}(2,1)}{a_c} \frac{f_s \cdot t_{\text{stance}}}{r^2 \cdot t_{\text{stride}}}$$

$$= \left[\left(\frac{\text{RRW} \cdot \text{ARW}}{a_c} \right)^2 + \frac{2\sigma_v^2 \text{RRW}^3}{a_c \cdot g} \sqrt{\frac{f_s t_{\text{stance}}}{r^2 t_{\text{stride}}}} \right]^{\frac{1}{2}}. \tag{6.44}$$

The uncertainty of gyroscope bias along the North is $\sigma_{g_N} = \sqrt{P_{44}(1,1)}$.
Attitude estimation covariance is obtained by combining (6.7) and (6.23)

$$-P_{11}(1,1) \cdot g \cdot t_{\text{stride}} = \frac{P_{12}(1,2)P_{22}(2,2)}{r^2} f_s \cdot t_{\text{stance}}, \tag{6.45}$$

and therefore $P_{11}(1,1)$ can be expressed as

$$P_{11}(1,1) = -\frac{P_{12}(1,2)P_{22}(2,2)}{g} \frac{f_s \cdot t_{\text{stance}}}{r^2 \cdot t_{\text{stride}}}. \tag{6.46}$$

The uncertainty of the attitude estimation along the North is $\sigma_\theta = \sqrt{P_{11}(1,1)}$.

To estimate the position uncertainty, we analyze the propagation of $P_{12}(3,2)$ first. Equation (6.11) shows that the propagation of $P_{12}(3,2)$ is related to the acceleration along the North a_N and the azimuth angle uncertainty $P_{11}(3,3)$. In a single gait cycle, $P_{11}(3,3)$ can be considered as constant since the duration of one gait cycle is relatively short (around one second). Thus, $P_{12}(3,2)$ is an integral of a_N, i.e. the real velocity of IMU along the North $v_N(t)$. Therefore, $P_{12}(3,2)$ returns to a near-zero value when the update step begins, and as a result, the update step has little effect on $P_{12}(3,2)$ since its value is already close to zero. $P_{12}(3,2)$ can be expressed as

$$P_{12}(3,2) \approx -P_{11}(3,3) \cdot v_N(t) = -\left(\text{ARW}^2 t + \frac{\text{RRW}^2}{3} t^3 \right) \cdot v_N(t). \tag{6.47}$$

Similarly, several other terms that are necessary in the derivation can be calculated as

$$P_{41}(3,3) = -\int_0^t P_{44}(3,3) \cdot \cos\theta \cdot d\tau = -\frac{\text{RRW}^2}{2} \cdot a_c \cdot t^2,$$

$$P_{41}(1,3) = \int_0^t P_{44}(1,1) \cdot \sin\theta \cdot d\tau = \sigma_{g_N}^2 \cdot a_s \cdot t,$$

$$P_{42}(1,2) = -\int_0^t P_{41}(1,3) \cdot a_N \cdot d\tau = -\sigma_{g_N}^2 \cdot a_s \cdot t \cdot v_N(t),$$

$$P_{42}(3,2) = -\int_0^t P_{41}(3,3) \cdot a_N \cdot d\tau = \frac{\text{RRW}^2}{2} \cdot a_c \cdot t^2 \cdot v_N(t),$$

$$P_{43}(1,2) = \int_0^t P_{42}(1,2) \cdot d\tau = -\int_0^t \sigma_{g_N}^2 \cdot a_s \cdot t \cdot v_N(t) \cdot d\tau$$

$$\approx -\sum_i \sigma_{g_N}^2 \cdot a_s \cdot t_i \int_{\text{cycle } i} v_N(t) \cdot d\tau$$

$$= -\sum_i \sigma_{g_N}^2 \cdot a_s \cdot t_i s_N = -\frac{1}{2} \sigma_{g_N}^2 \cdot a_s \cdot t^2 s_N,$$

$$P_{43}(3,2) = \int_0^t P_{42}(3,2) \cdot d\tau = \frac{RRW^2}{6} \cdot a_c \cdot s_N \cdot t^3,$$

$$P_{13}(3,2) = \int_0^t [P_{12}(3,2) + P_{43}(1,2)\sin\theta - P_{43}(3,2)\cos\theta]d\tau$$

$$= -\left(\frac{ARW^2}{2}t^2 + \frac{RRW^2}{12}t^4\right) \cdot s_N - \frac{1}{6}\sigma_{g_N}^2 \cdot a_s^2 \cdot s_N \cdot t^3$$

$$- \frac{RRW^2}{24} \cdot a_c^2 \cdot s_N \cdot t^4,$$

where a_s is the average value of $\sin\theta$ over the whole gait cycle, and s_N is the stride length of a human gait. In the equation for $P_{43}(1,2)$, the integral over the whole navigation process is calculated as the summation of the integral over each gait cycle i. In the integral of each gait cycle, t is approximated as a constant t_i and moved out of the integral because the relative changing rate of v_N is much larger than that of t in a single gait cycle.

Then, we estimate the level of P_{23} since it is related to the propagation of P_{33}, both in the predict step (indicated by (6.8)) and in the update step (indicated by (6.27) and (6.29)).

From our observation of many cases, the value of $P_{13}(2,1)$ remains at a constant level during the navigation. Therefore, a combination of (6.12) and (6.26) yields

$$P_{12}(2,1)t_{stride} = N_{stance}P_{23}(1,1) \cdot P_{12}(2,1)/r^2, \tag{6.48}$$

or equivalently represented as

$$P_{23}(1,1) = \frac{r^2 \cdot t_{stride}}{N_{stance}} = \frac{r^2 \cdot t_{stride}}{f_s \cdot t_{stance}}. \tag{6.49}$$

Comparing (6.9) and (6.10) yields

$$P_{23}(2,2) = P_{23}(1,1) - \int_0^t a_N P_{13}(3,2) \cdot d\tau \approx \frac{r^2 \cdot t_{stride}}{f_s \cdot t_{stance}} - P_{13}(3,2) \cdot v_N(t). \tag{6.50}$$

Now we explain the reason why the acceleration caused by foot motion cannot be neglected in (6.9) and (6.10). The position uncertainty $P_{33}(2,2)$ is derived by integrating $P_{23}(2,2)$ twice, and the acceleration term a_N will be transformed to the displacement term s_N. Therefore, even though the velocity v_N returns to zero after a full gait cycle, its integral, displacement s_N, cannot be neglected. The acceleration term in (6.6), however, is only integrated once to obtain the final result for velocity uncertainty. As a result, neglecting the acceleration term will not introduce large errors, but will only lose information about some fluctuations within a single gait cycle.

Combining (6.8) and (6.27), we can obtain the full increment of $P_{33}(1,1)$, which corresponds to the square of position uncertainty along the trajectory, during a complete gait cycle

$$\Delta P_{33}(1,1) = 2P_{23}(1,1) \cdot t_{\text{stride}} - \frac{P_{23}(1,1)^2}{r^2} \cdot N_{\text{stance}}$$

$$\approx \left(2 - \frac{t_{\text{stride}}}{4}\right) \frac{r^2 \cdot t_{\text{stride}}}{f_s \cdot t_{\text{stance}}} \cdot t_{\text{stride}}. \tag{6.51}$$

Therefore, the propagation of $P_{33}(1,1)$ can be expressed as

$$P_{33}(1,1) = \left(2 - \frac{t_{\text{stride}}}{4}\right) \frac{r^2 \cdot t_{\text{stride}}}{f_s \cdot t_{\text{stance}}} \cdot t. \tag{6.52}$$

The propagation of $P_{33}(2,2)$ can be derived similarly as

$$P_{33}(2,2) = \left(2 - \frac{t_{\text{stride}}}{4}\right) \frac{t_{\text{stride}} r^2}{t_{\text{stance}} f_s} \cdot t + \frac{1}{3} \text{ARW}^2 s_N^2 \cdot t^3 + \frac{a_s^2}{12} \sigma_{g_N}^2 s_N^2 \cdot t^4$$

$$+ \left(\frac{1}{30} + \frac{a_c^2}{60}\right) \text{RRW}^2 s_N^2 \cdot t^5. \tag{6.53}$$

We set $\sigma_{\parallel} = \sqrt{P_{33}(1,1)}$ and $\sigma_{\perp} = \sqrt{P_{33}(2,2)}$, where $\sigma_{\parallel}$ and $\sigma_{\perp}$ are the position estimation uncertainties parallel and perpendicular to the trajectory, and they correspond to 1.2 times of the semi-major and semi-minor axes of Circular Error Probable (CEP), respectively.

Equations (6.34), (6.40), (6.46), (6.52), and (6.53) fully describe the uncertainty of navigation results due to IMU noises with respect to the orientation, velocity, and position.

6.2.5 Observations

1. Angle Random Walk (ARW), Velocity Random Walk (VRW), and RRW all affect the final navigation uncertainties, for example in (6.40); higher noise level results in larger errors.
2. The velocity measurement uncertainty r plays an important role in the final results; lower r indicates a higher reliability and weight of the zero-velocity information in the EKF, resulting in a better navigation accuracy. However, this value is determined by human gait pattern and the type of floor [10]. Therefore, it should be adjusted according to the experiment and cannot be set arbitrarily.
3. The position uncertainty along the trajectory is dominated by the velocity measurement uncertainty r in the EKF and is proportional to square root of the navigation time.
4. The position uncertainty perpendicular to the trajectory depends on many parameters (see (6.53)). However, it is dominated by RRW and is proportional to the navigation time of the power of 2.5, in the case of long-term navigation.

5. Human gait pattern affects the navigation errors. It is reflected in the ratio between duration of the stance phase and the whole gait cycle and the average of sine and cosine value of the pitch angle, for example in (6.38). A higher percentage of the stance phase during the gait cycle gives the EKF more measurements to compensate for the IMU noises and reduced the overall navigation errors.

6. AcRW is not included in the model. This is due to our assumption that the propagated velocity covariance during the swing phase is much smaller than the velocity measurement uncertainty r. This conclusion agrees with the argument in [9].

7. The results are only approximations of the navigation errors due to assumptions and approximations made during the derivation, e.g. 2D foot motion, moderate IMU performance, high IMU sampling rate, and straight trajectory. Validity of the approximations will be demonstrated in the following section.

6.3 Verification of Analysis

It is instructive to consider some concrete examples to support the previously derived equations. In this section, we present verification of the analysis both numerically and experimentally.

6.3.1 Numerical Verification

First, we use computer simulation to verify the derived analytical expressions. In this example, first, a trajectory of foot toward the North and the corresponding IMU readouts are generated based on a human gait analysis reported in Section 6.1. Then, the numerical results are compared to analytical expressions. The generated trajectory is a straight line toward the North containing 100 steps. The total time duration of the trajectory is 53.6 seconds and the total length of the trajectory is 77 m.

6.3.1.1 Effect of ARW

We first present the influence of the ARW of gyroscopes on the navigation errors. We sweep the ARW value from 0.01 to 10 $°/\sqrt{h}$ (from near navigation grade to consumer grade), while keeping other parameters constant. VRW of accelerometers is set to be 0.14 $mg/\sqrt{Hz}$ (industrial grade), RRW of gyroscopes is set to be 0.048 $°/s/\sqrt{h}$, and the sampling frequency is selected to be 800 Hz. The simulation results are presented in Figure 6.6. The upper plot shows a relation between the ARW and the velocity estimation uncertainty and the lower plot shows a relation

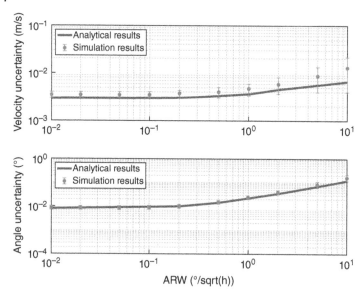

Figure 6.6 Effects of ARW of the gyroscopes on the velocity and angle estimation errors in the ZUPT-aided inertial navigation algorithm.

between the ARW and the angle estimation uncertainty. Notice that the angle estimation uncertainty is only for the roll and pitch angle, since the yaw angle error is unobservable in the EKF and propagates according to (6.34). In both plots, In both plots, the solid lines are analytical results, and the error bars are simulation results. The simulation results are a range instead of a value because covariances of the estimation errors fluctuate during the navigation (see Figure 6.5). The upper and lower bounds of the error bars show the amplitude of fluctuation and the squares show an average value of the fluctuation.

A close match between the analytical and simulation results verifies validity of the analysis. Figure 6.6 shows that both velocity estimation uncertainty and angle estimation uncertainty are not affected by ARW when its value is smaller than $0.1\ °/\sqrt{h}$. An explanation is that in this case the navigation uncertainty is dominated by other errors, such as VRW and RRW, therefore, it is independent of ARW value. The lower bound of the fluctuation of velocity uncertainty is almost not affected by ARW either. This is because the lower bound of the velocity uncertainty is limited by the velocity measurement uncertainty set in the EKF, which is fixed in this model. It is also noticed that fluctuation of the angle uncertainty is much smaller than the velocity uncertainty. The reason is that the velocity is directly observable in the ZUPT-aided navigation algorithm, and therefore, the EKF can directly estimate the velocity value and reduce the velocity uncertainty. The angle

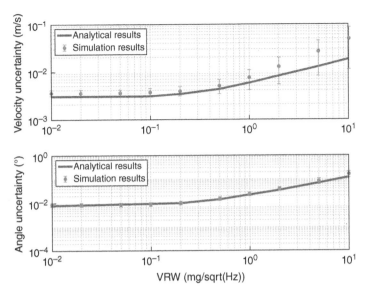

Figure 6.7 Effects of VRW of the accelerometers on the velocity and angle estimation errors in the ZUPT-aided inertial navigation algorithm.

estimation, however, is achieved through coupling the velocity and angle, and as a result, the observability is reduced.

6.3.1.2 Effect of VRW

Similarly, we sweep the VRW value of accelerometers from 0.01 to $10 \, \text{mg}/\sqrt{\text{Hz}}$, while keeping ARW of the gyroscope to be 0.21 $°/\sqrt{\text{h}}$ (industrial grade) and RRW to be 0.048 $°/\text{s}/\sqrt{\text{h}}$. The results are shown in Figure 6.7. As expected, the curves become flat when VRW is small, since the navigation error is dominated by gyroscope errors in this range.

6.3.1.3 Effect of RRW

As indicated in (6.53), RRW is the major error source that affects the navigation accuracy. We sweep the RRW value of gyroscopes from 6×10^{-4} to 0.6 $°/\text{s}/\sqrt{\text{h}}$, while keeping ARW of the gyroscope to be 0.21 $°/\sqrt{\text{h}}$ and VRW to be $0.14 \, \text{mg}/\sqrt{\text{Hz}}$. The influence of RRW on the velocity and angle estimation errors is shown in Figure 6.8.

Figure 6.9 shows the relation between the position uncertainty and RRW. A difference within 10% is demonstrated between the analytical results and the numerical results. Note that the position uncertainty perpendicular to the trajectory is not affected by RRW, but dominated by velocity uncertainty during the stance phase. As a result, a lower velocity measurement uncertainty is desirable for a

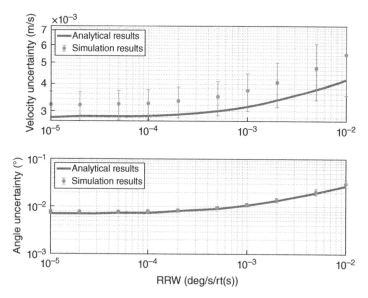

Figure 6.8 Effects of RRW of the gyroscopes on the velocity and angle estimation errors in the ZUPT-aided inertial navigation algorithm.

better navigation accuracy. The following considerations can help to reduce the velocity measurement uncertainty and improve the overall navigation accuracy:

1. A stiffer shoe with less deformation during walking.
2. A better position to attach the IMU, so that the state of the IMU can be closer to stationary during the stance phase.
3. Shock absorber on the shoes to prevent strong shocks between the shoe and the ground [11].

6.3.2 Experimental Verification

Next, we present an experimental example to verify the results.

In the experiment, a VectorNav VN-200 INS (Industrial grade) is mounted on the right shoe by a 3D-printed fixture (Figure 6.11) and IMU readouts are collected during walking. Allan deviations of the accelerometer and gyroscope readouts are collected to confirm the performance of the IMU [12], and the result is shown in Figure 6.10. ARW, VRW, and RRW of the IMU are 0.21 $°/\sqrt{h}$, 0.14 mg/$\sqrt{Hz}$, and 0.048 $°/s\sqrt{h}$, respectively. Sampling frequency is set to be 800 Hz (maximum sampling rate of the IMU) to capture the high-frequency component of the motion. The length of the straight-line trajectory is around 100 m, and the total navigation time is around 110 seconds. During the first 10 seconds of each run, the foot is

Figure 6.9 Relation between RRW of gyroscopes and the position estimation uncertainties.

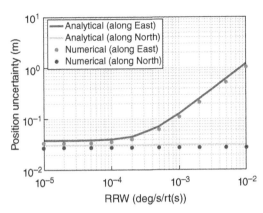

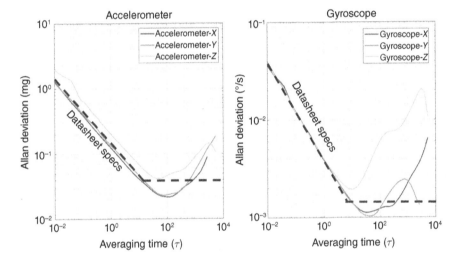

Figure 6.10 Allan deviation plot of the IMU used in this study. The result is compared to the datasheet specs [13].

stationary for the initial calibration of IMU and calculation of the initial roll and pitch angles. Magnetometer is used to determine the initial orientation of the foot. IMU data for 40 trajectories are collected to obtain a relatively accurate position uncertainty during the navigation.

The navigation error results of 40 trajectories are shown in Figure 6.11. In all cases, the nominal trajectory is a straight-line trajectory toward the North. All estimated trajectories exhibit a drift to the right side, and the averaged drift value is 1.82 m. This phenomenon is a result of the systematic error due to the nonzero velocity of the foot during the stance phase and the gyroscope g-sensitivity, and it

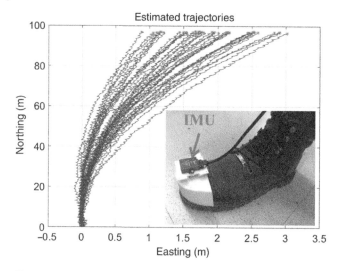

Figure 6.11 The navigation error results of 40 trajectories. The averaged time duration is about 110 seconds, including the initial calibration. Note that scales for the two axes are different to highlight the effect of error accumulation. Source: Wang et al. [11].

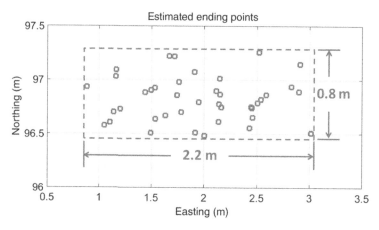

Figure 6.12 Ending points of 40 trajectories. All data points are in a rectangular area with the length of 2.2 m and width of 0.8 m.

has been reported in [9]. This drift is the result of a systematic errors, and it will be discussed in detail in Chapter 7.

A zoomed-in view of the ending points of 40 trajectories is shown in Figure 6.12. The ending points are distributed in a rectangular area with the length of 2.2 m and width of 0.8 m. Based on IMU performances from Figure 6.10 and analytical results from (6.52) and (6.53), the uncertainty of position estimation can be calculated as $\sigma_\parallel = 0.07$ m and $\sigma_\perp = 0.43$ m. Assuming the position error is normally distributed, then 99% of the points should be in an ellipse with the major axis of $6\sigma_\perp = 2.58$ m and the minor axis of $6\sigma_\parallel = 0.42$m. The analytical expression is within 20% of the experimental result along the direction perpendicular to the trajectory, showing a good agreement. As for the direction along the trajectory, the analytical result is about 50% smaller than the experimental result, possibly due to the systematic modeling errors, or the velocity uncertainty during the stance phases. The shock and vibrations of the IMU when the foot touches the ground are not considered in the analytical model, and they may also introduce extra errors. Similar phenomenon of larger navigation errors with IMU mounted on the foot has been reported previously in [14].

6.4 Limitations of the ZUPT Aiding Technique

ZUPT-aided inertial navigation algorithm eliminates the velocity drift during pedestrian navigation, and therefore greatly reduces the overall navigation error compared to the navigation result without any aiding. However, there are also a few limitations of the algorithm that need to be addressed in order to further improve the navigation accuracy of ZUPT-aided pedestrian inertial navigation [9]. In this section, we will mainly discuss the limitations of ZUPT, and more methods to mitigate the limitations will be discussed in detail in Chapter 7.

The most fundamental issue of ZUPT is that it assumes the velocity of the foot during the stance phase to be zero. As it has been reported in literature, such as [10, 15], the residual velocity of the foot during the stance phase is not zero. The residual velocity in each step will accumulate into a relatively large navigation error. Both underestimation of the stride length and the heading angle drift are the results of it. On the one hand, ZUPT is needed during each stance phase to eliminate the velocity drift; on the other hand, too much usage of ZUPT will introduce extra error due to the zero-velocity assumption. One mitigation method is to improve the ZUPT detector to optimize the usage of ZUPT during each step. Another method is to model the motion of the foot in a finer way, instead of simply assuming zero-velocity state. Extra sensors, such as pressure sensors, optical trackers, and magnetometers, can also be used to measure the motion of the foot and calibrate for any error caused by ZUPT.

Another limitation of the ZUPT aiding is that it requires the IMU to be placed close to the foot. Due to human gait dynamics, the motions of feet typically are of much larger amplitude than that of the body, with maximum angular rate of around 1000 to 2000 °/s and maximum specific force of around 10 to 15 g. Therefore, it requires a larger IMU measuring range than typical IMUs implemented

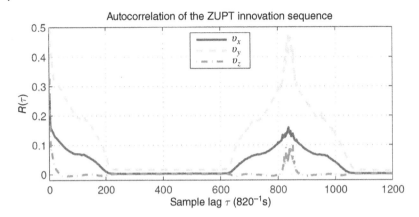

Figure 6.13 Autocorrelations of the *x*, *y*, and *z* components of the innovation sequence during ZUPT-aided pedestrian inertial navigation. Source: Nilsson et al. [9].

for vehicle navigation. Besides, due to the fast-changing specific force during the heel strike, a large sampling rate and bandwidth are also necessary, otherwise, systematic measurement error will be accumulated.

The algorithm also involves the detection of the ZUPT state, which is the process of a binary hypothesis test, where the stance phase and the swing phase will be detected by IMU readouts. In a hypothesis test problem, the distribution of the IMU readouts is needed for both stance phase and swing phase, which is related to many parameters, such as walking pattern, step pace, and floor type. However, in a typical navigation problem, none of the parameters are known. Therefore, the parameters for the detector are usually empirically tuned in an ad-hoc manner, which limits the compatibility of the algorithm. Adaptive detectors based on Bayesian approach and machine learning method can identify some of the important parameters and partially solve this issue.

The ZUPT aiding uses the EKF to merge the information from the IMU and that from the ZUPT. However, the EKF assumes unbiased, uncorrelated, and normal process noise and measurement noise, which is often not the case in the ZUPT-aided pedestrian inertial navigation. As shown in Figure 6.13, the innovations during the ZUPT-aided pedestrian inertial navigation are correlated. The correlation may introduce extra systematic errors in the estimation.

6.5 Conclusions

In this chapter, an analytical model correlating the IMU errors and the navigation errors in the ZUPT-aided pedestrian inertial navigation was presented. Both numerical and experimental examples were given to illustrate the effects of different parameters. The analytical model matches the numerical simulation and experimental results with errors of 10% and 20%, respectively. The drift of the estimated orientation of the trajectory due to the RRW of the z-axis gyroscope is found to be the main factor that affects the navigation accuracy. We believe that a relatively small discrepancy between the analytical result and the numerical simulation is indicative of the accuracy of the analysis, while a relatively larger discrepancy between the analytical result and the experiment is likely the result of systematic modeling errors, such as biased and correlated process noises, nonlinear navigation dynamics, and the zero-velocity assumption. This will be addressed in Chapter 7.

This chapter estimates the magnitude of navigation errors due to IMU errors, laying a basis for error analysis in pedestrian inertial navigation. It is envisioned to aid in analysis of the effect of errors in sensors, which might lead to a well-informed selection of sensors for the task of ZUPT-aided pedestrian inertial navigation.

References

1 Tao, W., Liu, T., Zheng, R., and Feng, H. (2012). Gait analysis using wearable sensors. *Sensors* 12 (2): 2255–2283.

2 Murray, M.P., Drought, A.B., and Kory, R.C. (1964). Walking gait of normal man. *Journal of Bone & Joint Surgery* 46: 335–360.

3 Perry, J. and Davids, J.R. (1992). Gait analysis: normal and pathological function. *Journal of Pediatric Orthopaedics* 12 (6): 815.

4 Whittle, M.W. (2002). *Gait Analysis: An Introduction*, 3e. Oxford: Butterworth-Heinemann.

5 Wang, Y., Chernyshoff, A., and Shkel, A.M. (2018). Error analysis of ZUPT-aided pedestrian inertial navigation. *IEEE International Conference on Indoor Positioning and Indoor Navigation (IPIN)*, Nantes, France (24–27 September 2018).

6 Jimenez, A.R., Seco, F., Prieto, J.C., and Guevara, J. (2010). Indoor pedestrian navigation using an INS/EKF framework for yaw drift reduction and a foot-mounted IMU. *IEEE Workshop on Positioning Navigation and Communication (WPNC)*, Dresden, Germany (11–12 March 2010).

7 Wang, Y., Vatanparvar, D., Chernyshoff, A., and Shkel, A.M. (2018). Analytical closed-form estimation of position error on ZUPT-augmented pedestrian inertial navigation. *IEEE Sensors Letters* 2 (4): 1–4.

8 Titterton, D. and Weston, J. (2004). *Strapdown Inertial Navigation Technology*, 2e, vol. 207. AIAA.

9 Nilsson, J.O., Skog, I., and Handel, P. (2012). A note on the limitations of ZUPTs and the implications on sensor error modeling. *IEEE International Conference on Indoor Positioning and Indoor Navigation (IPIN)*, Sydney, Australia (13–15 November 2012).

10 Wang, Y., Askari, S., and Shkel, A.M. (2019). Study on mounting position of IMU for better accuracy of ZUPT-aided pedestrian inertial navigation. *IEEE International Symposium on Inertial Sensors & Systems*, Naples, FL, USA (1–5 April 2019).

11 Wang, Y., Chernyshoff, A., and Shkel, A.M. (2020). Study on estimation errors in ZUPT-aided pedestrian inertial navigation due to IMU noises. *IEEE Transactions on Aerospace and Electronic Systems* 56 (3): 2280–2291.

12 El-Sheimy, N., Hou, H., and Niu, X. (2008). Analysis and modeling of inertial sensors using Allan variance. *IEEE Transactions on Instrumentation and Measurement* 57 (1): 140–149.

13 VectorNav (2020). VN-200 GPS-Aided Inertial Navigation System Product Brief. https://www.vectornav.com/docs/default-source/documentation/vn-200-documentation/PB-12-0003.pdf?sfvrsn=749ee6b9_13.

14 Laverne, M., George, M., Lord, D. et al. (2011). Experimental validation of foot to foot range measurements in pedestrian tracking. *ION GNSS Conference*, Portland, OR, USA (19–23 September 2011).

15 Peruzzi, A., Della Croce, U., and Cereatti, A. (2011). Estimation of stride length in level walking using an inertial measurement unit attached to the foot: a validation of the zero velocity assumption during stance. *Journal of Biomechanics* 44 (10): 1991–1994.

7

Navigation Error Reduction in the ZUPT-Aided Pedestrian Inertial Navigation

Many error sources contribute to the overall navigation error in the Zero-Velocity Update (ZUPT)-aided pedestrian inertial navigation. They can generally be categorized into two groups: errors caused by the Inertial Measurement Unit (IMU) and errors caused by the navigation algorithm. As mentioned in Chapter 4, the IMU errors can be divided into stochastic type and systematic, or constant type. Typical stochastic IMU errors include sensor white noise and walking bias. The walking bias can be estimated and compensated by the Extended Kalman Filter (EKF), while the white noise can only be reduced by utilizing higher grade of IMU. The systematic errors are relatively constant, and they can be compensated by calibration before the navigation.

On the other hand, errors caused by the navigation algorithm mainly come from the inaccurate modeling used in the algorithm. For example, as discussed in Chapter 6, the velocity of the foot may not be exactly zero during the stance phase, while a zero-velocity assumption is used in the algorithm. The algorithm-caused navigation errors can also be categorized into stochastic type and systematic type. The navigation error sources are summarized in Table 7.1. Note that the residual velocity may contain both the stochastic and systematic components.

In this chapter, we will discuss a few methods that can be implemented in the ZUPT-aided pedestrian inertial navigation in order to reduce the navigation errors. We limit the methods introduced in this chapter such that they only require the foot-mounted IMU, and no extra sensing modality is necessary. Other methods involving more sensing modalities will be covered in Chapter 9.

Pedestrian Inertial Navigation with Self-Contained Aiding, First Edition. Yusheng Wang and Andrei M. Shkel.
© 2021 The Institute of Electrical and Electronics Engineers, Inc. Published 2021 by John Wiley & Sons, Inc.

Table 7.1 Possible error sources in the ZUPT-aided pedestrian inertial navigation.

Possible error types		Error source	
		IMU-related	**Algorithm-related**
Error property	Stochastic	White noise	Correlated noise
		Walking bias	Residual velocity
	Systematic	Nonorthogonality	Stance phase mis-detection
		G-sensitivity	Residual velocity

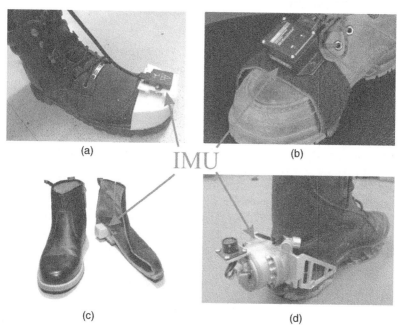

IMU

(a)   (b)

(c)   (d)

Figure 7.1 Possible IMU-mounting positions. Source: Refs. [1–4].

7.1 IMU-Mounting Position Selection

The ZUPT-aided pedestrian inertial navigation requires a foot-mounted IMU to collect data. Many different IMU-mounting locations have been explored, as shown in Figure 7.1. Here, we are attempting to answer the following questions: "which part of the foot is the optimal IMU-mounting position?" "How do we define 'optimal'?" Some qualitative conclusions have been made in Chapter 6, such as lower shock level and longer stance phase. In this section, we attempt to answer these questions quantitatively by comparing the two most commonly used IMU-mounting positions: on top of the forefoot and behind the heel.

7.1.1 Data Collection

Two identical industrial grade IMUs (VectorNav VN-200 IMU) are rigidly mounted above the forefoot and behind the heel of the boot, respectively, to collect data of the motion of the forefoot and the heel simultaneously. Noise characteristics of the two IMUs are first estimated by Allan deviation analysis and compared to the datasheet [5]. The results are presented in Figure 7.2, showing that the two IMUs have the same noise level, eliminating some possible discrepancies in experiment due to IMU characteristics.

With IMUs mounted on the forefoot and behind the heel of the boot, experiments are conducted by different subjects, each of which walks on different floor types at different speeds, assuring generality and validity of conclusions. For each experiment, a trajectory with 600 strides (1200 steps) is recorded. The walking pace is fixed with the help of a metronome for a better postprocessing result, but the step counter is not used in estimation of the navigation accuracy. Floor types, such as hard floor, grass lawn, and sand floor, are tested. Walking paces range from 84 to 112 steps per minute. Trajectories, such as walking upstairs and downstairs, are also investigated. Four different walking patterns by four different subjects are tested in the study. IMU readouts are collected during walking and then analyzed in postprocessing.

7.1.2 Data Averaging

Collected IMU data from the forefoot and the heel are processed to compare the two mounting positions. IMU data are first averaged to reduce the IMU noise and extract parameters, such as length of the stance phase and the shock level during walking.

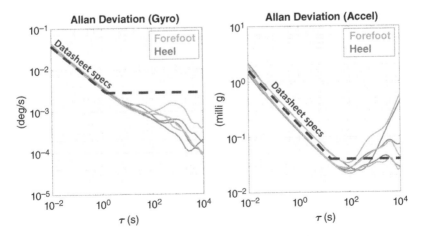

Figure 7.2 Noise characteristics of the IMUs used in the study.

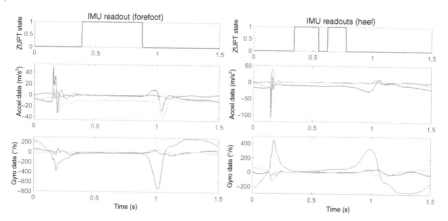

Figure 7.3 Comparison of averaged IMU data and ZUPT states from IMUs mounted on the forefoot and behind the heel. Stance phase is identified when ZUPT state is equal to 1.

In this process, IMU data of 600 gait cycles are averaged. The main purpose of averaging is to remove the majority of the white noise for a better extraction of motion features. A zero-velocity detector is applied to the averaged data to determine the stance phase during the gait cycle. The results are shown in Figure 7.3. On the left are the ZUPT state and the averaged IMU data from the forefoot. On the right of Figure 7.3 are the readouts of the IMU located behind the heel. ZUPT state is derived for both mounting positions with the same threshold. For IMUs mounted on the forefoot and behind the heel, Figure 7.3 shows an average time duration of the stance phase of 0.498 and 0.363 seconds, and the shock experienced by IMU is on the level of 80 and 150 m/s², respectively. Both a longer stance phase and a lower level of shock yield a better navigation accuracy for the IMU mounted on the forefoot. Interruption of the stance phase for the heel shows that the IMU is moving during the stance phase, indicating a less stable stance phase if IMU is mounted behind the heel. One disadvantage of forefoot as the IMU mounting position is that the maximum gyroscope readout is about 800 °/s compared to the 450 °/s when the IMU is mounted behind the heel. In the study, the gyroscopes of IMU have a maximum measuring range of 2000 °/s, and therefore in this case, the higher magnitude of gyroscope readout is not an issue. However, the choice of IMU with a sufficient measuring range is an important consideration for this application.

Bandwidth of most MEMS-based inertial sensors is typically not high enough for foot-mounted inertial navigation due to the shock during the heel-strike [6]. However, in this study, the bandwidth of the sensors is 250 Hz, and the effects of the limited bandwidth may be negligible [7].

Table 7.2 Stance phase analysis summary with different floor types.

Floor type	Step pace (step/min)	Velocity uncertainty (m/s)		Stance phase length (s)	
		Heel	Forefoot	Heel	Forefoot
Hard floor	84	0.022	0.016	0.36	0.50
	100	0.025	0.020	0.33	0.39
	112	0.029	0.024	0.29	0.34
Grass lawn	84	0.046	0.032	0.48	0.55
	100	0.052	0.035	0.38	0.45
	112	0.055	0.045	0.34	0.39
Sand	84	0.060	0.048	0.51	0.48
	100	0.076	0.050	0.37	0.37
	112	0.095	0.051	0.31	0.32

7.1.3 Data Processing Summary

The stance phase analysis with different step paces and floor types is presented in Table 7.2. The method to determine the velocity uncertainty of the foot during the stance phase has been introduced in Chapter 5, and thus, it will not be discussed here. The velocity uncertainty during the stance phase is highest when walking on the sand, then on the grass lawn, and it is lowest when walking on the hard floor. This result is expected, since more uncertainty will be induced when walking on softer surfaces. The same conclusion is drawn in all nine scenarios that the forefoot is a better IMU mounting position than the heel with an average of 20% lower velocity uncertainty and about 20% longer stance phase period.

Table 7.3 lists the analysis of walking upstairs and downstairs, while Table 7.4 lists the analysis of walking on a hard floor by different subjects (different walking patterns). A similar result is obtained for all cases, concluding that a longer stance phase and a lower velocity uncertainty should be expected to be achieved with IMU mounted on the forefoot.

All the values listed in Tables 7.2–7.4 can be used as a guidance when determining the velocity uncertainty during the stance phase. Some conclusions can be drawn from the statistics:

- Floor types affect the velocity uncertainty during the stance phase the most among factors considered in the study. The harder the floor is, the smaller the velocity uncertainty will be. For example, in the case, where the IMU is mounted on the forefoot and the step pace is 84 steps per minute, the velocity uncertainty is 0.016 m/s when walking on the hard floor, while the value increases

Table 7.3 Stance phase analysis summary with different trajectories.

Trajectory	Step pace (step/min)	Velocity uncertainty (m/s)		Stance phase length (s)	
		Heel	Forefoot	Heel	Forefoot
Upstairs	84	0.086	0.042	0.51	0.53
	100	0.088	0.038	0.39	0.45
	112	0.086	0.029	0.33	0.39
Downstairs	84	0.084	0.055	0.48	0.58
	100	0.083	0.042	0.31	0.42
	112	0.080	0.038	0.27	0.30

Table 7.4 Stance phase analysis summary with different subjects.

Subject number	Step pace (step/min)	Velocity uncertainty (m/s)		Stance phase length (s)	
		Heel	Forefoot	Heel	Forefoot
Subject 1	84	0.022	0.016	0.36	0.50
	100	0.025	0.020	0.33	0.39
	112	0.029	0.024	0.29	0.34
Subject 2	84	0.044	0.028	0.49	0.50
	100	0.040	0.026	0.29	0.31
	112	0.034	0.020	0.27	0.27
Subject 3	84	0.052	0.035	0.48	0.52
	100	0.071	0.026	0.34	0.37
	112	0.022	0.020	0.28	0.30
Subject 4	84	0.029	0.026	0.38	0.42
	100	0.040	0.034	0.28	0.33
	112	0.031	0.021	0.27	0.28

to 0.032 m/s when walking on the grass lawn, and the value reaches 0.048 m/s if walking on the sand. No obvious relation between the floor type and the averaged stance phase length can be concluded.

- In both cases of walking upstairs and downstairs, the velocity uncertainty during the stance phase is much lower with the IMU mounted on the forefoot than on the heel. It can be explained as follows: most people mainly use forefoot as support when walking either upstairs or downstairs, while the heel is mostly in the air. The solid support from the stair to the forefoot results in a much lower velocity uncertainty. Besides, the time during the stance phase is also longer with IMU mounted on the forefoot.

- Results for four agents in total have been reported in this study. The gait patterns of four of them will be different, but the conclusion can be made that the velocity uncertainty will be lower and the stance phase will be longer with IMU mounted on the forefoot.
- Higher step pace results in shorter stance phase duration. However, the relation between the step pace and the velocity uncertainty during the stance phase is not clear. It might be related to the gait patterns of different people. On the one hand, slower step pace results in a more solid step on the floor, which reduces the motion of the foot.
- Stance phase duration is longer if IMU is mounted on the forefoot. It can be explained by the duration of each phase in the gait cycles shown in Figure 6.2. There is about 40% of the whole gait cycle when the heel is on the floor, while the percentage is about 45% for the forefoot. As for the velocity uncertainty during the stance phase, forefoot is also preferable with a lower velocity uncertainty in all different scenarios.

7.1.4 Experimental Verification

Next, a direct correlation between the IMU mounting position and the navigation accuracy is presented to compare the forefoot and the heel as IMU mounting positions.

A circular path with a diameter of 8 m and 10 laps is used as a trajectory to experimentally demonstrate the effect of the IMU mounting position on the Circular Error Probable (CEP). The trajectory is a close loop to facilitate the extraction of navigation position errors. Thirty-four tests are recorded with an average navigation time of 260 seconds, and the navigation errors for all tests are presented in Figure 7.4. The CEP is reduced from 1.79 to 0.96 m by using the data from the IMU mounted on the forefoot instead of behind the heel.

A comparison of estimation results from IMU mounted at the forefoot and the heel is presented in Figure 7.5. It can be clearly observed that the estimated trajectory is smoother with IMU mounted at the forefoot, and it is due to a smaller position correction during the stance phase, indicating less noise accumulated during the swing phases. To better interpret the performance of the EKF, we characterize the innovation of the EKF. Innovation, or measurement residual, is defined as the difference between the measurement and its prediction [8]. In the case of the ZUPT-aided pedestrian inertial navigation, innovation is the difference between estimated velocity of the foot during the stance phases and the pseudo-measurement of the velocity of the foot. The distribution of the

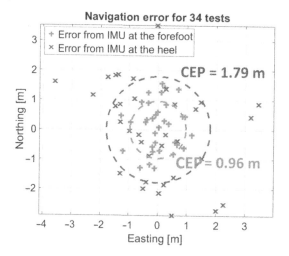

Figure 7.4 Navigation error of 34 tests of the same circular trajectory.

innovation from the same experiment is shown in the lower part of Figure 7.5. A lower innovation covariance is shown in the data from the IMU mounted at the forefoot, indicating a better navigation accuracy. Note that some innovation values are outside of 3σ envelop, and they are probably related to the false alarm in the ZUPT detection.

7.2 Residual Velocity Calibration

It has been demonstrated that underestimation of the trajectory length is related to the zero-velocity assumption of the foot during the stance phase [6]. To quantitatively analyze the relation between the zero-velocity assumption and the underestimate of the trajectory length, the motion of the foot during the stance phase needs to be recorded and analyzed.

In the exemplary experiment to analyze the motion of the foot, a magnetic motion tracking system can be used. In this experiment, there are two parts in the system: a magnetic source placed on the floor as a reference, and a tracker mounted by a custom fixture on top of the IMU (VectorNav VN-200). The experimental setup is shown in Figure 7.6. The tracking system is able to track the relative position between the tracker and the source with a nominal resolution of 1mm at a sampling frequency of 60 Hz [9]. Velocity of the foot is derived by taking derivative of the relative position with respect to time.

Seventy gait cycles are recorded with a walking pace of approximately 84 steps per minute toward the North, and the results are shown in Figure 7.7. The thick

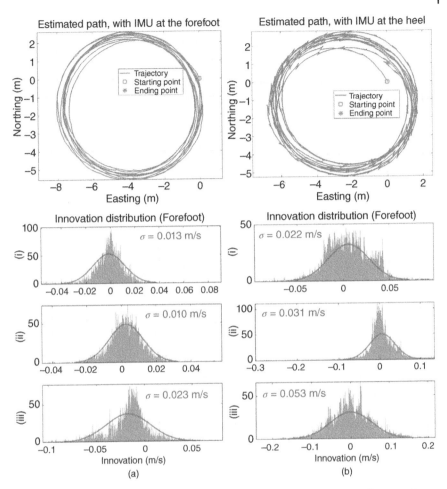

Figure 7.5 Comparison of estimated trajectories and innovations from IMU mounted at the forefoot (a) and the heel (b).

solid lines are the averaged velocities along three directions. The stance phase corresponds approximately to the time period between 0.5 and 0.9 seconds (indicated by the gray box), and the velocity of the foot is close to zero during the stance phase. However, Figure 7.8 shows the zoomed-in view of the velocity during the stance phase. The thick solid lines are the averaged velocities of the foot, and the dashed lines in light gray correspond to zero velocity. A residual velocity on the order of 0.01 m/s is clearly observed, and therefore, assuming the nonzero residual velocity during the stance phase to be zero will introduce a systematic error. The dashed lines in dark gray are the 1σ interval of the velocity distribution,

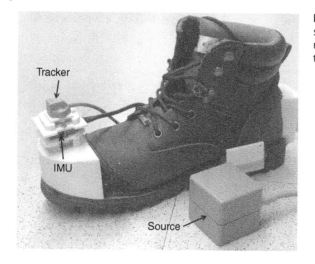

Figure 7.6 Experimental setup to record the motion of the foot during the stance phase.

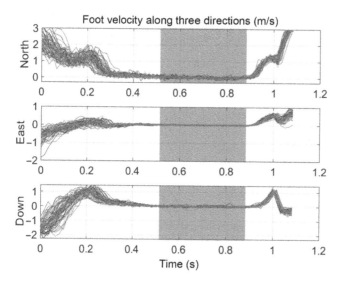

Figure 7.7 Velocity of the foot along three directions during a gait cycle. The thick solid lines are the averaged velocities along three directions.

showing a velocity uncertainty about 0.02 m/s during the stance phase. The noisy individual measurement is due to the relatively low sampling rate of the motion tracker and the derivative operation to extract the velocity from the distance. Note that the fluctuation caused by the derivative operation will increase the measured velocity variance. Therefore, the velocity variance derived in this method cannot

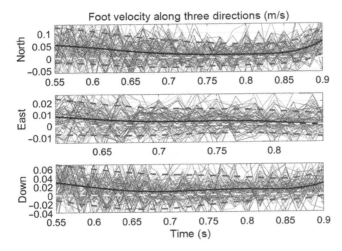

Figure 7.8 Zoomed-in view of the velocity of the foot during the stance phase. The light gray dashed lines correspond to zero-velocity state, and the dark gray dashed lines are the 1σ range of the velocity distribution.

be used to determine the velocity uncertainty. However, the measured mean value of the velocity can still be considered accurate.

Underestimate of the trajectory length is directly related to the residual velocity during the stance phase. However, the residual velocity is not constant during the stance phase. Therefore, its average value is related to the length of the stance phase as determined by the ZUPT detector. Figure 7.9 shows the test statistics of the same 70 steps recorded previously and the thick solid line is an averaged value.

The level of the test statistic is not a constant even when the foot is in the stance phase. Therefore, the length of the detected stance phase is related to the value of the predefined threshold. For example, the detected stance phase is between 0.65 and 0.8 s, if the threshold is set to 1×10^4 (the dashed green line in Figure 7.9). If the threshold is increased to 3×10^4, the detected stance phase is between 0.57 and 0.86 seconds (the outer dashed line in Figure 7.9). A longer detected stance phase lead to a higher averaged residual velocity of the foot during the stance phase, and therefore, a higher systematic error will be introduced.

To experimentally verify the effects of residual velocity during the stance phase, IMU data are recorded for 10 straight trajectories with length of 100 m. The walking pace is 84 steps per minute. For each of the trajectory, thresholds ranging from 1×10^4 to 5×10^4 are applied in the ZUPT detector, and the underestimate of the trajectory length is recorded and shown in Figure 7.10. The thick solid line is the result of the previous analysis, and the thinner lines are experimental results. A good match is demonstrated, verifying that the residual velocity during the stance phase is the major factor that lead to the underestimate of the trajectory length.

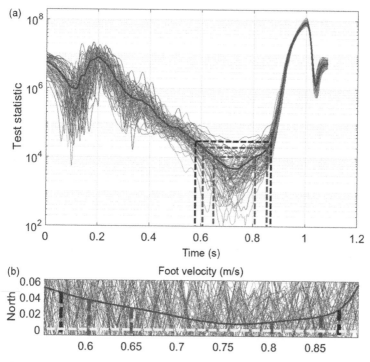

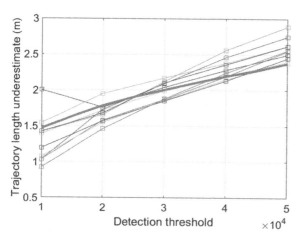

Figure 7.9 Panel (a) shows the test statistics of the same 70 steps recorded previously. Thick solid line is an averaged value. Panel (b) shows the residual velocity of the foot along the trajectory during the stance phase. The inner, middle, and outer dashed lines correspond to threshold levels of 1×10^4, 2×10^4, and 3×10^4, respectively.

Figure 7.10 Relation between the underestimate of trajectory length and the ZUPT detection threshold. The thick solid line is the result of the previous analysis, and the thinner lines are experimental results from 10 different runs.

7.3   Gyroscope G-Sensitivity Calibration

Gyroscope g-sensitivity is the erroneous measurement of a gyroscope in response to the external acceleration. Due to the high shock that the IMU experiences during human walking, the effect of the gyroscope g-sensitivity cannot be neglected.

Trajectory orientation drift in the ZUPT-aided pedestrian inertial navigation is believed to be related to the g-sensitivity of gyroscopes [10, 11]. Due to a severe dynamics of the foot during walking, the heading angle error is demonstrated to accumulate at a rate of 135 °/h in [12], even though the gyroscope bias instability is only 3 °/h.

To relate the trajectory orientation drift and the heading angle drift, IMU data are recorded for a straight-line trajectory of 550 m toward the North. The experimentally recorded trajectory is shown by the solid line in Figure 7.11, showing a drift to the right. The estimated heading angle is shown in the inset in Figure 7.11, showing a drift at the rate of 0.028 °/s. The dashed line shows an analytically generated trajectory assuming a constant speed and a heading angle increase at the rate of 0.028 °/s. The experimental and generated trajectories match each other with a difference within 10 m, indicating that the heading angle drift is the major factor that leads to the trajectory orientation drift.

To understand the reason for the heading angle drift, the IMU is full calibrated to obtain not only the gyroscope and accelerometer biases, but also the non-orthogonality and gyroscope g-sensitivity. Such data are not typically provided for consumer or industrial grade IMUs. During calibration, the IMU is rigidly mounted on a tilt table to achieve different orientations, and the tilt table is mounted on a single-axis rate table (IDEAL AEROSMITH 1270VS) to generate a constant reference rotation. The experimental setup is shown in Figure 7.12a.

Figure 7.11   The solid line is an estimated trajectory, and the dashed line is an analytically generated trajectory with heading angle increasing at a rate of 0.028 °/s. Note that the scales for the x and y axes are different. The inset shows that the estimated heading angle increases at a rate of 0.028 °/s.

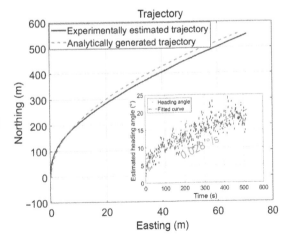

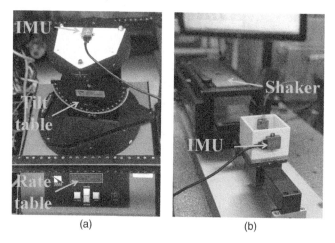

(a) (b)

Figure 7.12 (a) Experimental setup to statically calibrate IMU; (b) experimental setup to measure the relation between gyroscope g-sensitivity and acceleration frequency [13].

A standard IMU calibration procedure is followed [14], and the calibration results are as follows

$$
b_a = \begin{bmatrix} -0.025 \\ -0.0176 \\ 0.1955 \end{bmatrix}, \quad
M_a = \begin{bmatrix} 1.0020 & -0.0083 & -0.0042 \\ 0.0055 & 0.9986 & 0.0051 \\ 0.0067 & -0.0039 & 0.9964 \end{bmatrix},
$$

$$
b_g = \begin{bmatrix} -0.0893 \\ 0.0375 \\ -0.0412 \end{bmatrix}, \quad
M_g = \begin{bmatrix} 0.9972 & -0.0041 & -0.0067 \\ 0.0041 & 0.9972 & 0.0052 \\ 0.0067 & -0.0027 & 1.0019 \end{bmatrix},
$$

$$
G_g = \begin{bmatrix} 0.0041 & 0.0002 & -0.0005 \\ 0.0002 & 0.0025 & 0.0002 \\ -0.0005 & -0.0006 & -0.0022 \end{bmatrix},
$$

where b_a is the accelerometer bias in m/s^2, b_g is the gyroscope bias in $^\circ$/s, M_a is the accelerometer misalignment matrix, M_g is the gyroscope misalignment matrix, and G_g is the gyroscope g-sensitivity matrix in $^\circ$/s/(m/s^2). These values are used in the IMU readout compensation before fed into the navigation algorithm in the following studies. Note that the g-sensitivity of the gyroscope is on the order of 0.002 $^\circ$/s/(m/s^2). A shock on the order of 10 g, which is typical for a foot-mounted IMU, will cause a gyroscope bias of 0.2 $^\circ$/s/(m/s^2) and resulted in a large navigation error, if not compensated.

Note that the g-sensitivity of gyroscopes is obtained in a static condition. Since the IMU will experience severe dynamics during navigation, a measurement of the gyroscope g-sensitivity in dynamic conditions is also necessary. To achieve

Figure 7.13 Relation between the gyroscope g-sensitivity and the vibration frequency obtained from three independent measurements. The dashed line is the gyroscope g-sensitivity measured in static calibration. Inset is the FFT of the z-axis accelerometer readout during a typical walking of two minutes.

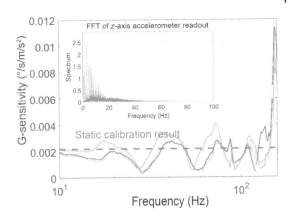

it, the IMU is rigidly mounted on a shaker (APS Dynamics APS-500), and the gyroscope readouts are recorded while the shaker generated vibrations with different frequencies, ranging from 10 to 160 Hz. The experimental setup is shown in Figure 7.12b. Three independent measurements are conducted to guarantee repeatability. A relation between the gyroscope g-sensitivity and the vibration frequency is shown in Figure 7.13. Gyroscope g-sensitivity remains relatively stable around 0.0022 °/s/(m/s²) until the vibration frequency is above 140 Hz. Inset is the FFT of the z-axis accelerometer readout during a typical walking of two minutes, and the spectrum reaches close to zero with frequencies higher than 80 Hz. Therefore, the gyroscope g-sensitivity can be considered constant for the whole frequency range in the case of pedestrian navigation.

7.4 Navigation Error Compensation Results

In this section, the effects of residual velocity and gyroscope g-sensitivity calibrations on the navigation error are experimentally demonstrated.

Two steps are identified as necessary to compensate for systematic errors identified: (i) calibrate the IMU readouts to remove the effects of sensor biases, nonorthogonality, and especially the gyroscope g-sensitivity; (ii) set the pseudo-measurement of the velocity of the foot during the stance phase according to the gait pattern, which is characterized in Section 7.2, instead of zero.

In the experimental setup used for illustration, the IMU is mounted on top of the toes. A straight-line trajectory of 99.6 m is used, and 40 sets of data are recorded in total. The navigation results with and without the compensation are shown in Figure 7.14. It can be seen that the drift in trajectory orientation is compensated, while the compensation effects along the trajectory cannot be seen clearly due to the scale. Figure 7.15 shows the ending points of the 40 trajectories with

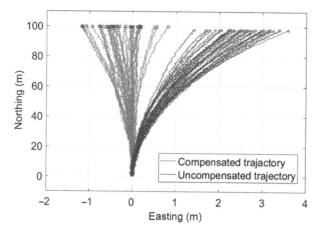

Figure 7.14 Comparison of trajectories with and without systematic error compensation. Note that the scales for *x* and *y* axes are different.

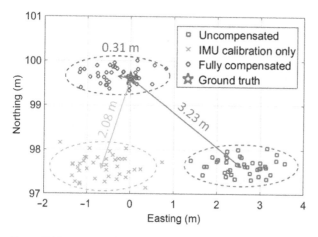

Figure 7.15 Comparison of the end points with and without systematic error compensation. The dashed lines are the 3σ boundaries of the results.

and without compensation. The dashed lines are the 3σ boundaries of the results. Note that they are approximately of the same size, since they are the result of stochastic noise, which is not compensated in this section. Note that the navigation error demonstrated in Figure 7.15 agrees with the model developed in Chapter 6, indicating that IMU noise is the dominant navigation error source. The averaged navigation error is 3.23 m without any compensation (gray squares in Figure 7.15). The majority part of the navigation error perpendicular to the trajectory can be canceled by calibrating the IMU, and the averaged navigation error is reduced to

2.08 m (light gray crosses in Figure 7.15). After implementing the residual velocity compensation, the navigation error along the trajectory is compensated, and the averaged navigation error is reduced to 0.31 m (gray circles in Figure 7.15), demonstrating a more than 10× improvement. After systematic error compensation, the error caused by IMU noises becomes dominant, thus requiring the improvement in the IMU performance to improve the overall navigation accuracy.

7.5 Conclusions

In this chapter, we presented some methods to modify the standard ZUPT-aided pedestrian inertial navigation in order to further reduce the navigation error. No extra sensing modality is needed in these methods. More specifically, mounting the IMU on the forefoot instead of on the heel is shown to reduce the stochastic error, or equivalently CEP, by 50%. Compensation of residual velocity during the stance phase and gyroscope g-sensitivity are demonstrated to reduce the systematic error over 10×.

Methods reported in this chapter eliminate the majority of the errors caused by the ZUPT implementation, which would otherwise be the dominant error source in the ZUPT-aided pedestrian inertial navigation. The IMU error becomes dominant factor in the navigation error after implementing all the compensation methods discussed in this chapter, providing the possibility of further reducing the navigation error by improving the performance of inertial sensors. Therefore, the results reported in this chapter can serve as a foundation for further pedestrian navigation system development, including improvement of inertial sensor performance, and addition of other aiding mechanisms to the system, such as altimeter, ultrasonic ranging, signals of opportunities, and cooperative localization. They will be discussed in Chapters 9 and 10.

References

1 Wang, Y. and Shkel, A.M. (2019). Adaptive threshold for zero-velocity detector in ZUPT-aided pedestrian inertial navigation. *IEEE Sensors Letters* 3 (11): 1–4.

2 Bird, J. and Arden, D. (2011). Indoor navigation with foot-mounted strapdown inertial navigation and magnetic sensors [emerging opportunities for localization and tracking]. *IEEE Wireless Communications* 18 (2): 28–35.

3 Nilsson, J.-O., Gupta, A.K., and Händel, P. (2014). Foot-mounted inertial navigation made easy. *IEEE International Conference on Indoor Positioning and Indoor Navigation (IPIN)*, Busan, South Korea (27–30 October 2014), pp. 24–29.

4 Laverne, M., George, M., Lord, D. et al. (2011). Experimental validation of foot to foot range measurements in pedestrian tracking. *ION GNSS Conference,* Portland, OR, USA (19–23 September 2011), pp. 1386–1393.

5 VectorNav (2020). VN-200 GPS-Aided Inertial Navigation System Product Brief. https://www.vectornav.com/docs/default-source/documentation/vn-200-documentation/PB-12-0003.pdf?sfvrsn=749ee6b9_13.

6 Nilsson, J.O., Skog, I., and Handel, P. (2012). A note on the limitations of ZUPTs and the implications on sensor error modeling. *IEEE International Conference on Indoor Positioning and Indoor Navigation (IPIN),* Sydney, Australia (13–15 November 2012).

7 Wang, Y. and Shkel, A.M. (2020). A review on ZUPT-aided pedestrian inertial navigation. *27th Saint Petersburg International Conference on Integrated Navigation Systems,* Saint Petersburg, Russia (25–27 May 2020).

8 Bar-Shalom, Y., Li, X.-R., and Kirubarajan, T. (2001). *Estimation with Applications to Tracking and Navigation: Theory Algorithms and Software.* Wiley.

9 Polhemus (2017). PATRIOT two-sensor 6-DOF tracker. https://polhemus.com/_assets/img/PATRIOT_brochure.pdf (accessed 08 March 2021).

10 Bancroft, J.B. and Lachapelle, G. (2012). Estimating MEMS gyroscope g-sensitivity errors in foot mounted navigation. *IEEE Ubiquitous Positioning, Indoor Navigation, and Location Based Service (UPINLBS),* Helsinki, Finland (3–4 October 2012).

11 Zhu, Z. and Wang, S. (2018). A novel step length estimator based on foot-mounted MEMS sensors. *Sensors* 18 (12): 4447.

12 Laverne, M., George, M., Lord, D. et al. (2011). Experimental validation of foot to foot range measurements in pedestrian tracking. *ION GNSS Conference,* Portland, OR, USA (19–23 September 2011).

13 Wang, Y., Lin, Y., Askari, S., Jao, C., and Shkel, A.M. (2020). Compensation of systematic errors in ZUPT-Aided pedestrian inertial navigation. *2020 IEEE/ION Position,* Location and Navigation Symposium (PLANS), pp. 1452–1456.

14 Chatfield, A.B. (1997). *Fundamentals of High Accuracy Inertial Navigation.* American Institute of Aeronautics and Astronautics. eISBN: 978-1-60086-646-3.

8

Adaptive ZUPT-Aided Pedestrian Inertial Navigation

In pedestrian inertial navigation, the navigation system may be used under various navigation scenarios. For example, the subject may move with different patterns: walking, running, jumping, and even crawling; the subject may move on different types of floors: going upstairs, going downstairs, going uphills, and going down-hills; and the subject may even work on the floors of different materials: concrete, sand, and grass. Different paces of motion will also lead to changes in the human gait dynamics. All these different factors may affect the parameters to be used in the Zero-Velocity Update (ZUPT)-aided pedestrian inertial navigation algorithm, such as the threshold used for the stance phase detection, the residual velocity of the foot during the stance phase, and the corresponding measurement uncertainty. In order to guarantee the navigation accuracy in different scenarios, the parameters should be adjusted accordingly, i.e. the navigation algorithm should be able to adapt to different navigation scenarios. In this chapter, we report approaches for the adaptive ZUPT-aided pedestrian inertial navigation.

8.1 Floor Type Detection

Due to the complicated dynamics of the foot during navigation, many methods have been developed to adapt the navigation system to various navigation scenarios. For example, floor type is an important factor to be determined. However, to the best of our knowledge, the floor type detection is typically used in vacuum cleaners based on pressure sensors [1], ultrasonic transducers [2], and motor power monitors [3]. In this section, we present an Inertial Measurement Unit (IMU)-based floor type detection for pedestrian inertial navigation [4]. In this method, one foot-mounted IMU is required in the system as before, which greatly reduces the complexity of the overall system. After conducting the floor

Pedestrian Inertial Navigation with Self-Contained Aiding, First Edition. Yusheng Wang and Andrei M. Shkel.
© 2021 The Institute of Electrical and Electronics Engineers, Inc. Published 2021 by John Wiley & Sons, Inc.

type detection, the classification result is then used in the ZUPT-aided pedestrian inertial navigation in the multiple-model approach, and the improvement of the navigation accuracy demonstrates the effects of the floor type detection.

8.1.1 Algorithm Overview

The adaptive algorithm includes four main steps, and its schematics is presented in Figure 8.1. First, the IMU data are divided into different partitions of length M, with each partition corresponding to a full gait cycle. The length of each partition is fixed, so that they are of the same dimension to facilitate the following steps. The IMU readout of each partition is used as an input for the following floor type identification. Second, the Principal Component Analysis (PCA) is conducted to the input data to reduce their dimensionality from $6M$ to p. The value p can be

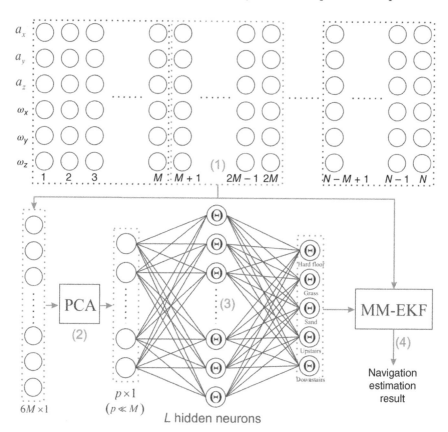

Figure 8.1 Schematics of the algorithm discussed in this chapter. The numbers (1)–(4) indicate the four main steps in the algorithm.

selected by considering a trade-off between the identification accuracy and the computational load: more dimensionality reduction by the PCA leads to a simpler learning step at the cost of lower identification accuracy. Third, a two-layer Artificial Neural Network (ANN) is trained to conduct the floor type identification. The number of neurons in the hidden layer L should be properly selected, where another trade-off between the accuracy and the computational load is involved. Fourth, in our proposed implementation, the floor type identification result is used in a Multiple Model Extended Kalman Filter (MM-EKF) for the ZUPT-aided pedestrian inertial navigation.

8.1.2 Algorithm Implementation

In this section, we describe some details of the algorithm and the trade-offs involved in the parameter selection.

8.1.2.1 Data Partition

The IMU readouts are first partitioned into different gait cycles (shown in Figure 8.2). One of the most recognizable features is the y-axis gyroscope peak during the toe-off of the foot, which is used as the start mark of each partition. In this study, the gait frequency is approximately 90 steps per minute and the IMU sampling rate is 400 Hz, and thus, the length of each partition is set to be 533. Notice that the actual length of each gait might be different, indicating that each partition may not necessarily end at the following toe-off event. For example, there is a gap between the second and third data partition in Figure 8.2, indicating

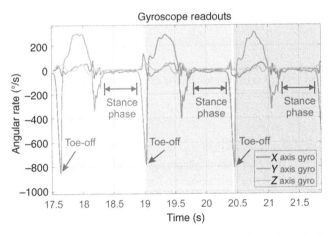

Figure 8.2 An example of IMU data partition. Each partition (indicated by different brightness) starts at toe-off of the foot.

that the second data partition ends before the third partition begins. However, as it will be shown later, this is not a problem in this problem formulation.

8.1.2.2 Principal Component Analysis

The IMU readout at each time step contains six measurement data, including three from accelerometer readouts and three from gyroscope readouts. In this example, the dimensionality of each partition is $533 \times 6 = 3198$, which may lead to large amount of computation in the following training of the neural network, if no dimensionality reduction is performed. Therefore, it is necessary to conduct the PCA to reduce the dimensionality of data before the following learning steps [5].

The PCA is one of the most commonly used technique for reducing the dimensionality of large datasets, in order to increase interpretability of the data while minimizing information loss [6]. The basic idea of the PCA is to find new variables, or principal components, that independent linear functions of the original dataset and successively maximize the variance. It can be shown that the PCA can be accomplished by solving an eigenvalue/eigenvector problem. More specifically, if a dataset containing N numerical variables of dimension d is given, it can be represented by a $d \times N$ data matrix

$$X = \begin{bmatrix} x_1 & x_2 & \cdots & x_N \end{bmatrix},$$

whose ith column corresponds to the ith data x_i. Instead of the original data matrix, it is common to use the centered data matrix

$$X_c = \begin{bmatrix} x_1 - \bar{x} & x_2 - \bar{x} & \cdots & x_N - \bar{x}, \end{bmatrix},$$

where

$$\bar{x} = \frac{1}{N} \sum_{i=1}^{N} x_i$$

is the sample mean. Then, the Singular Value Decomposition (SVD) can be applied to the centered data matrix

$$X_c = USV^{\mathrm{T}}, \tag{8.1}$$

where U and V are $d \times r$ and $N \times r$ matrices with orthonormal columns, which are called the left and right singular vectors, respectively, with $r = \min(d, N)$, and S is a diagonal matrix whose diagonal terms are the singular values of the centered data matrix X_c.

Next, let U_p be the first p columns of U, corresponding to the p largest singular values of X_c. Then, the PCA feature vectors are given by

$$y_i = U_p^{\mathrm{T}}(x_i - \bar{x}), \quad i = 1, 2, \ldots, N. \tag{8.2}$$

Figure 8.3 Distribution of eigenvalues of the centered data matrix after conducting the singular value decomposition.

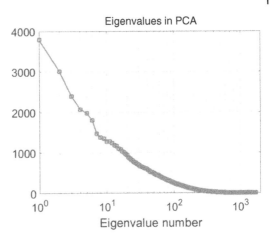

Notice that the size of y_i is $p \times 1$, and thus, the dimensionality of the data is reduced from d to p.

In this example, 1673 partitions are first collected and labeled with the corresponding floor types: walking on hard floor, walking on grass, walking on sand, walking upstairs, and walking downstairs. The corresponding eigenvalues of the centered data matrix after applying the SVD are shown in Figure 8.3. The eigenvalues are distributed continuously, and no threshold can be obtained by observation. Therefore, the validation approach is needed to determine the optimal output dimension of the PCA.

8.1.2.3 Artificial Neural Network

After applying the PCA to reduce the dimensionality, a two-layer ANN is trained and used for the floor type identification. Backpropagation algorithm with mini-batch gradient descent is used to train the weights in the ANN [7]. However, the number of neurons in the hidden layer L should be determined before training the neural network.

To sum up, there are two parameters to be selected in total: the dimensionality of the PCA output p and the number of hidden neurons L; both of them will affect the complexity and accuracy of the algorithm. Validation method is used to determine the appropriate values for the two parameters. About 20% of training data are randomly selected and reserved as an independent validation set to evaluate the training effects. The rest 80% of data are utilized to train the neural network. The optimal weights of the ANN are selected to be those that achieve highest identification accuracy over the validation set. Notice that the output of the neural network is the probability of each class. For simplicity, the class with the highest probability is selected as the classification result in this study.

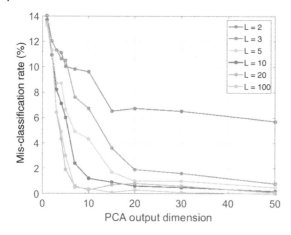

Figure 8.4 Relation between the misclassification rate, PCA output dimension, and number of neurons in the hidden layer.

Figure 8.4 presents the misclassification rate of the algorithm over the entire available data with different parameter selections. The x-axis represents the PCA output dimension, and the y-axis represents the misclassification rate. Different lines show the results with different number of neurons in the hidden layer. Generally, higher PCA output dimension p and larger number of neurons in the hidden layer improve the performance. If only two neurons are used in the hidden layer (indicated by the highest solid line), the misclassification rate is always higher than 5% regardless of the PCA output dimension. However, if the number of hidden neurons is increased to 3 (the second highest solid line), the misclassification rate can be reduced to around 1% with the PCA output dimension of over 50. When the number of hidden neurons is greater than 20 (the second lowest solid line), the improvement in the misclassification rate due to increasing number of neurons is only marginal. As a result, the parameter L is fixed to be 20 from this point. Notice that the mini-batch gradient descent is a stochastic gradient descent method in its nature, and therefore, small fluctuation in Figure 8.4 is expected.

On the other hand, the PCA output dimension p also affects the accuracy of classification. Figure 8.5 shows the confusion matrices of the floor type identification results with the PCA output dimension of 3 and 10, respectively. A classification accuracy of 93.9% is demonstrated with $p = 3$. The accuracy is improved to 99.5% as p increased to 10, with only a few misclassifications occurring between walking on the grass and walking on the sand, and between walking upstairs and downstairs. The misclassification can be explained as follows:

- The floor is soft for the case of sand and grass, and the hardness of the floor may vary depending on the thickness of the sand or grass layer even within the same floor type.
- The motion of the foot is more random when it is on the floor in the case of walking on sand and grass compared to the case of hard floor. Therefore, more outliers may occur and become misclassified.

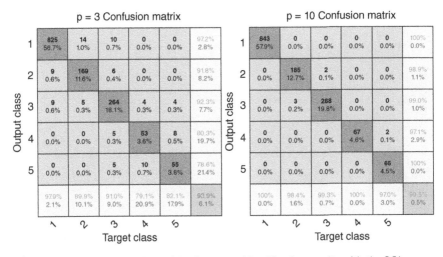

Figure 8.5 Confusion matrices of the floor type identification results with the PCA output dimension of 3 and 10, respectively. Classes are (1) walking on hard floor, (2) walking on grass, (3) walking on sand, (4) walking upstairs, and (5) walking downstairs.

- The gait patterns of walking upstairs and downstairs are generally different from the case of walking on the flat surface. Therefore, it is less likely to misclassify between the first three classes and the last two.
- The foot is less stable when it is on the stairs due to a large magnitude of motion. Thus, it is possible to mis-classify between them and walking on sand, especially if the PCA output dimension p is small, and much information is lost after applying PCA to the original data.

Notice that in the case of $p = 10$, high classification accuracy is obtained even with dimensionality reduction of more than 300 times. This indicates the effectiveness of the PCA method in reducing the dimensionality of data while keeping the key features.

Figure 8.6 presents the distribution of the first and second principal components. It shows that walking on the hard floor can be classified from other patterns even only with the first two principal components. However, walking on the grass and sand, and walking upstairs and downstairs are still indistinguishable from each other, and more principal components are needed to classify them.

8.1.2.4 Multiple Model EKF
The multiple-model approach was developed for estimation problems with parameters that may switch within a finite set of values [8, 9], and pedestrian inertial navigation on different floor types is exactly within the scope. Therefore, the MM-EKF is used to adapt the algorithm to different floor types. In the standard multiple-model Kalman filter approach, a set of parallel Kalman filters (KFs) are involved, each representing a model with the different parameters.

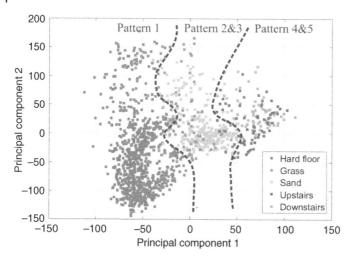

Figure 8.6 Distribution of the first two principal components of the available data.

The Kalman filter provides both the estimated system state and the innovation, or the measurement residual, of the estimate. The innovations from all Kalman filters are input into a hypothesis test to calculate the probability of each model. The final estimation results are obtained by calculating the weighted average of the estimated results from all pairs. In this study, instead of estimating the probability of each model based on hypothesis and the innovation of each Kalman filter, the floor type detection discussed previously is used. The schematics of the algorithm used in this study is shown in Figure 8.7.

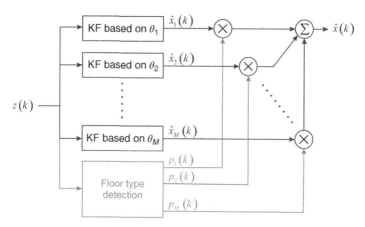

Figure 8.7 Schematics of the algorithm used in this study. The part in gray shows its difference from the standard multiple-model Kalman filter.

In this example of implementation, the differences between each model include the threshold for ZUPT detection, the residual velocity during the stance phase, and the corresponding velocity uncertainty. Each of these parameters can be obtained by separate calibrations, and the details have been reported in the previous chapters and will not be covered here.

8.1.3 Navigation Result

A navigation result is presented as an example showing the effect of the floor type detection on the improvement of navigation accuracy. In order to test the effect of the floor type identification, a trajectory involving sand, hard floor, and stairs is used. The subject first walks on the sand for about two minutes, then walks on the hard floor for another two minutes, and finally walks upstairs for about 1.5 minutes. The total length of the trajectory is about 320 m. The navigation results with and without the floor type identification are presented in Figure 8.8. The dashed line is the ground truth, and the solid line is the estimated trajectory without the floor type detection. The estimated trajectory with the floor type detection is presented by three segments, corresponding to three different floor

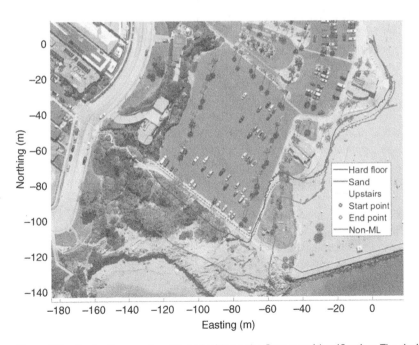

Figure 8.8 Navigation results with and without the floor type identification. The dashed line is the ground truth.

types identified by the algorithm. The smaller final estimation error verifies the effects of the floor type detection on improving the accuracy of pedestrian inertial navigation.

8.1.4 Summary

In this section, the algorithmic implementation of the floor type identification for pedestrian inertial navigation has been discussed. A combination of the PCA, ANN, and multiple-model Kalman filter is discussed as a potential adaptive implementation, showing that it is able to achieve a high classification accuracy with a reasonable amount of calculation. Floor type identification accuracy of as high as 99.5% is obtained over the training set, and a reduction of the navigation error is demonstrated, verifying the effects of the floor type identification.

The floor type identification is considered as the first essential step in the multiple-model Kalman filter to determine the parameters to be used. However, even when walking on the same floor type, many parameters to be used still will change with different gait frequencies. Therefore, it is also necessary to derive an adaptive stance phase detection, which is the subject Section 8.2.

8.2 Adaptive Stance Phase Detection

In a typical ZUPT detector, determination of the stance phase is conducted by comparing the test statistic to a threshold. However, different gait patterns lead to different gait dynamics even when walking on the same floor type, and as a result, different thresholds are required. The simplest way is to tune the threshold in an ad-hoc manner to achieve the best performance [10]. However, this is not practical in real navigation applications due to the lack of the ground truth in most navigation applications. Some other alternatives include adjusting the parameters based on the period of gait cycle (or equivalently, the walking speed) obtained by presetting the walking speed [11], and applying smoothed pseudo Wigner–Ville distribution (SPWVD) to the gyroscope readout to extract gait frequency [12]. Sensor fusion type approach is another option to adaptively detect the stance phases. For example, some studies report as an effective solution to mount the pressure sensors in the shoe sole to detect the pressure between the shoe and the floor [13], while others consider multiple IMUs to detect the motion of foot, shank, and thigh simultaneously to improve detection accuracy [14]. Machine learning-based stance phase detection has also been explored [15].

In this section, we present an adaptive threshold based on the Bayesian approach to enable the detector to work with various walking or running speeds [16]. This approach does not require extra sensing modalities and imposes less computational load on the system compared to the machine learning approaches.

8.2.1 Zero-Velocity Detector

A zero-velocity detector can be mathematically expressed as a binary hypothesis test, where the detector can choose between the two hypotheses: the IMU is moving (H_0) or the IMU is stationary (H_1). A common approach is to apply the Neyman–Pearson theorem and compare the likelihood ratio with some predefined threshold γ [17]: choose H_1 if

$$L(\mathbf{z}_n) = \frac{p(\mathbf{z}_n \mid H_1)}{p(\mathbf{z}_n \mid H_0)} > \gamma, \tag{8.3}$$

where $\mathbf{z}_k = \{\mathbf{y}_k\}_{k=n}^{n+N-1}$ is the N consecutive IMU readouts between time index n and $n + N - 1$, and $L(\cdot)$ is the likelihood ratio of probability of the measurement.

Stance Hypothesis Optimal dEtector (SHOE) is one of the most commonly used zero-velocity detectors [18]. This detector is based on the fact that the foot is almost stationary during the stance phase, and therefore, the magnitude of the specific force is equal to the gravity, and the angular rate is close to zero. However, many parameters that determine the Probability Density Functions (PDFs) of observations are unknown due to complicated dynamics of the human gait [19]. One possible solution is to replace the unknown parameters with their Maximum Likelihood (ML) estimates, and this method is called the Generalized Likelihood Ratio Test (GLRT) [20]. Using this method, the test statistic can be expressed as

$$L'_{\mathrm{ML}}(\mathbf{z}_n) = -\frac{2}{N} \log(L_{\mathrm{ML}}(\mathbf{z}_n)) = \frac{1}{N} \sum_{k=n}^{n+N-1} \frac{1}{\sigma_a^2} \left\| \mathbf{y}_k^a - g \frac{\bar{\mathbf{y}}^a}{\|\bar{\mathbf{y}}^a\|} \right\|^2 + \frac{1}{\sigma_\omega^2} \|\mathbf{y}_k^\omega\|^2, \tag{8.4}$$

where $\mathbf{y}_k^a$ and $\mathbf{y}_k^\omega$ are the accelerometer and the gyroscope readouts at time index k, respectively, $\bar{\mathbf{y}}^a$ is the averaged value of the N consecutive accelerometer readouts, σ_a and σ_ω are related to the white noise level of the accelerometer and the gyroscope, and g is the gravity.

We can then state the GLRT as: choose H_1 if

$$L'_{\mathrm{ML}}(\mathbf{z}_n) < \gamma', \tag{8.5}$$

where γ' is the threshold to be determined.

8.2.2 Adaptive Threshold Determination

For different gait patterns, the distributions of $\mathbf{z}_n$ are also different, and therefore, different thresholds are needed. In this section, we derive an adaptive threshold based on a time-dependent cost function. The main goal of the adaptive threshold is to adjust the ZUPT detector to different walking or running patterns, in order to minimize the extra navigation errors by erroneously applying ZUPT.

There are three general goals of adaptive threshold:

- Limit the probability of false alarm. False alarm happens if the detector determines the IMU is stationary while the foot is actually moving. False alarm will cause KF to erroneously set the velocity close to zero, which greatly degrades the results of navigation.
- Minimize the probability of miss-detection. Stance phase is the time period when zero-velocity information can be utilized to suppress the navigation error growth. Miss-detection will reduce the chance of compensation, therefore increase the overall navigation error.
- Adjust the threshold parameters automatically to fit different gait dynamics and maintain a proper amount of ZUPTs.

Figure 8.9 shows a typical test statistic $L'_{\mathrm{ML}}(z_n)$ for different gait dynamics. There are six different gait dynamics shown in Figure 8.9, corresponding to walking at the pace of 80, 90, 100, 110, and 120 steps per minute, and running at the pace of 160 steps per minute, respectively. High test statistic indicates that the IMU is moving while a lower value shows that the IMU is close to the stationary state. Figure 8.9 shows that the test statistic is around 50 when the foot is stationary, and this value is mostly related to IMU noises. The dark gray dashed lines are the averaged values of test statistic during the stance phase related to different gait dynamics, ranging from the lowest of 4×10^4 to the highest of 6×10^5. Note that

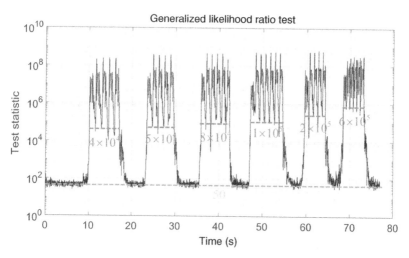

Figure 8.9 The solid line is a typical test statistic for different walking and running paces. The dark gray dashed lines show the test statistic levels during the stance phase with different gait paces, and the light gray dashed line shows the test statistic level when standing still.

these values are much higher than the test statistic when the foot is stationary on the floor, indicating that the foot is actually not stationary during the stance phase. Therefore, an excessive use of ZUPT will cause a degradation to the overall navigation accuracy [21]. Since the lowest values of the test statistic are different with varying walking or running speeds, an adaptive threshold is necessary to make the ZUPT detection more robust, especially in the case where a varying walking speed is involved.

The Bayesian likelihood ratio test states that the threshold can be expressed as

$$\gamma = \frac{p(H_0)}{p(H_1)} \cdot \frac{c_{10} - c_{00}}{c_{01} - c_{11}}, \tag{8.6}$$

where $p(H_0)/p(H_1)$ is the prior probability of the hypotheses, c_{00}, c_{11}, c_{10}, and c_{01} are the cost functions of correct detection of the swing phase, correct detection of the stance phase, a false alarm, and a miss-detection, respectively. We assume a uniform prior probability since no other information of the motion is available, and a zero cost for correct detections. Thus, the threshold equals the ratio of the cost function of the false alarm and miss-detection.

In a miss-detection, the stance phase of the foot is not detected, and therefore, the zero-velocity information is not fused in the system to suppress the error propagation. The associated cost is time-dependent, since the navigation error accumulates as a polynomial with respect to time without any error suppression [22]. Therefore, it is proper to assume a polynomial cost function for miss-detection instead of an exponential cost function used by some studies. On the other hand, the false alarm is the case where the zero-velocity information is fused into the system while the foot is still moving. The cost of a false alarm is related to the actual velocity of the foot, and it is relatively random and time-independent compared to the cost of a miss-detection [16]. Therefore, a constant cost function is assumed for the false alarm. To summarize, the ratio of the cost of a false alarm to the cost of a miss-detection can be expressed in a polynomial form

$$\gamma = \frac{c_{10}}{c_{01}} = \alpha_1 \cdot \Delta t^{-\theta_1}, \tag{8.7}$$

where Δt is the time difference between the previous ZUPT event and the current time step, and α_1 and θ_1 are design parameters to be decided. The threshold γ' can be defined as

$$\gamma' = -\frac{2}{N} \log(\gamma) = -\frac{2}{N}(\log(\alpha_1) - \theta_1 \cdot \log(\Delta t)) \triangleq \theta \cdot \log(\Delta t) + \alpha. \tag{8.8}$$

The value of γ' is low immediately after detection of the last stance phase, since the time interval Δt is small. The physical interpretation of this fact is that the probability of detecting another stance phase is low, according to (8.8), since we do not expect two stance phases to be very close to each other.

Another advantage of polynomial cost can be derived from (8.8). Note that (8.8) would be in the form of $\theta \cdot \Delta t + \alpha$, if an exponential cost were to be used. For normal gait patterns, the range of Δt is typically around one second. The slope of $\log(\Delta t)$, which is associated with the polynomial cost, is similar to that of Δt, which is in turn associated with the exponential cost, when Δt is around one second, but the slope is much larger if Δt is smaller than one second. Therefore, a similar performance of the stance phase detection can be expected with better robustness against false alarm in between the two stance phases.

The threshold γ' increases at a speed defined by θ as Δt increases, and α biases the overall level of the threshold. Ideally, θ should be defined such that γ' increases to the level of the test statistic during the stance phase in one gait cycle. This requires an estimation of the level of the test statistic during the stance phase, which is directly related to the gait frequency. In this study, we propose to take advantage of the shock level that the IMU experiences during the heel strike as an indicator to estimate the real-time gait frequency.

As the step pace increases, the minimum test statistic increases, as well as the shock level that the IMU experiences during the heel strike. The relation between the shock level and the minimum test statistic is shown in Figure 8.10. The dots correspond to data from different gait cycles, the solid line is the fitted curve, and the dashed lines are 1σ interval of the fitting. An exponential formula can be used to approximate the relation, and the parameter θ in (8.8) can be defined as

$$\theta = \epsilon \cdot \exp(0.0307 \times \text{Shock} + 8.6348), \tag{8.9}$$

where *Shock* is the shock level during the heel strike in (m/s), ϵ is the parameter that can be adjusted to achieve a proper length of the stance phase, and it is set to be 3.5 in this study. The parameter α adjusts the overall threshold level to reduce the probability of miss-detection and to improve the algorithm robustness.

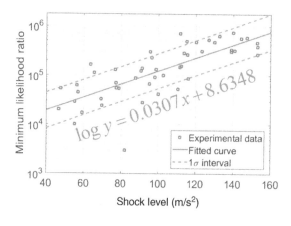

Figure 8.10 The relation between the shock level and the minimum test statistic in the same gait cycle. The dots correspond to data from different gait cycles, the solid line is a fitted curve, and the dashed lines are 1σ intervals.

Figure 8.11 The dashed lines in dark and light gray are adaptive thresholds with and without an artificial holding, respectively. The dots indicate the stance phases detected by the threshold without holding, while the stance phases detected by the threshold with holding is shown by the gray boxes.

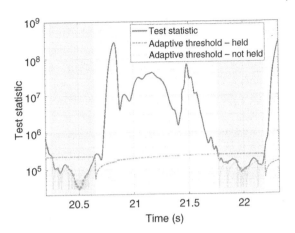

Advantages of using the shock level during the heel strike to extract gait frequency include:

- Priori knowledge of the gait frequency is not needed;
- Ability to continuously track the gait frequency with no lag;
- Less amount of computation than FFT or machine learning.

Note that the adaptive threshold in (8.8) is related to δt, indicating that the threshold will drop immediately after a stance phase is detected. In this case, only one time step of the stance phase can be determined by the detector (shown by the discrete dots in Figure 8.11). This will deteriorate the navigation accuracy since not all stance phase events are utilized. To ensure that enough ZUPTs can be implemented to suppress the navigation error, it is proposed to hold the threshold γ' until it becomes smaller than the test statistic again, instead of allowing it to drop back as (8.8) indicates. The effect of holding is shown in Figure 8.11. The threshold level remains constant during the stance phase (indicated by the gray boxes), which enables us to detect the whole stance phase, instead of discrete time instances during the stance phase.

8.2.3 Experimental Verification

Exemplary experiments are conducted to verify the effects of the adaptive threshold. A straight trajectory of 75 m is used in the experiment. The IMU is rigidly mounted at the forefoot and the sampling rate is 200 Hz. During the navigation, the subject first stands still for about 12 seconds, then walks at a pace of 84 steps per minute for about 15 seconds. It is followed by running at a pace of 160 steps per minute for about 15 seconds and walking for another 20 seconds. At last, the subject stands for about five seconds. The position propagation, accelerometer readouts, GLRT, and the navigation results are presented in Figure 8.12. Figure 8.12a shows the estimated position over time, and the velocity difference between walking and running can be clearly observed. Figure 8.12b shows the specific force of the IMU. A much higher shock level of about 150 m/s^2 during running is observed, exceeding the level of about 50 m/s^2 during walking. Figure 8.12c shows the test statistic (dark gray solid line) and the adaptive threshold (light gray dashed line). The threshold successfully captures the changes in dynamics during walking and running, and the threshold value is about 2×10^5 (light gray dashed line) and 2×10^6 (black dashed line), respectively. Figure 8.12d shows the comparison of estimated trajectories with different threshold settings. In the case of fixed threshold, the stance phase during running cannot be detected, if the threshold is set to be 2×10^5, which is proper for walking. Thus, the estimated trajectory drifts away soon after the subject started running. On the other hand, if the threshold is set to be 2×10^6 to detect the stance phase during running, too many ZUPTs are imposed during walking, and the estimated trajectory becomes much shorter than the ground truth by 12 m, corresponding to 16% of the total trajectory length. By applying adaptive threshold for the stance phase detection, the navigation error is reduced to 3 m. The effect of holding the threshold during the stance phase is shown by further reducing the navigation error from 3 to 1 m.

Navigation Root-Mean-Square Error (RMSE) has been analyzed with respect to different fixed thresholds for the same trajectory, and the result is presented in Figure 8.13. The minimum RMSE with a fixed threshold is 0.98 m, corresponding to a threshold 1.65×10^6. This value is between the adaptive threshold values for walking and running, and therefore, it can be considered as a trade-off: some stance phases during running can be detected, while not too much ZUPT will be imposed during walking. The RMSE for adaptive threshold is 0.61 m, and it is lower than the lowest RMSE achieved by any fixed threshold, showing the advantage of adaptive threshold for ZUPT detection. Note that the optimal fixed threshold is related to many parameters, such as walking speed, floor type, and walking pattern, and therefore, it is typically not available in most navigation scenarios, and can only be determined empirically. As a result, we expect that adaptive threshold will generally perform much better than any fixed threshold, especially in the case where walking or running paces are changing during navigation, as in the case shown in Figure 8.12.

8.2.4 Summary

To sum up, we present a method to set the threshold for the stance phase detection in the Bayesian approach, such that the ZUPT-aided pedestrian inertial navigation algorithm is adaptive to different gait frequencies. The method developed in this

Figure 8.12 Sub figures (a) through (d) show position propagation, specific force of the IMU, generalized likelihood ratio test, and the navigation results of the experiment, respectively. Note that the *x* and *y* axis scalings in (d) are different.

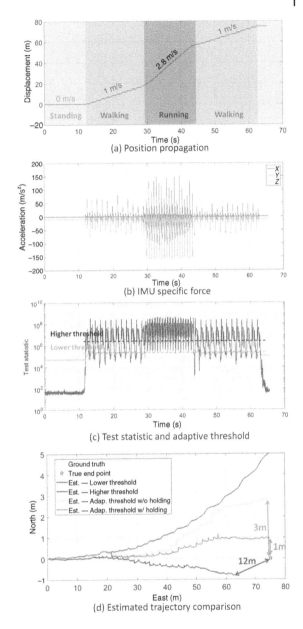

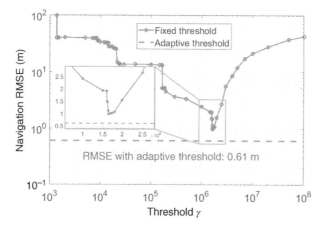

Figure 8.13 Relation between the navigation RMSE and fixed threshold level is shown by the solid line. The navigation RMSE achieved by adaptive threshold is shown by the dashed line.

section is shown to be adaptive to a wide range of moving speed, from as low as walking at 80 steps per minute to running at 160 steps per minute. Notice that the results shown in this section are only obtained in the case of moving on the hard floor as a demonstration; however, the results for other floor types can be obtained similarly.

8.3 Conclusions

Pedestrian inertial navigation system may work under various scenarios with different parameters to be selected. Therefore, an adaptive algorithm is needed to be incorporated with the ZUPT aiding to achieve the accuracy and robustness of the overall system. In this chapter, we mainly introduced approaches to enable the ZUPT-aided pedestrian inertial navigation to adapt to different floor types and gait frequencies. More than 10× of navigation error reduction has been demonstrated in both efforts for certain navigation trajectories. Notice that for the floor type detection, although it may require much computation to determine the neural network parameters during the training stage, the presentation stage is cheaper computationally. As for the adaptive threshold method, the threshold level can be directly correlated with the shock level, and thus not much extra computational power is needed. Therefore, both of these approaches can be implemented in real time.

There are still many open questions to be answered to further improve the adaptivity of the ZUPT-aided pedestrian inertial navigation. For example, it is still not clear how to choose appropriate models to be included in the multiple-model approach to cover as many scenarios as possible while not making the algorithm too complicated. A focused study is necessary on how to cover not only different floor types, but also different gait patterns.

References

1 Delmas, G. and Driessen, J.A.T. (1998). Vacuum cleaner with floor type detection means and motor power control as a function of the detected floor type. US Patent 5,722,109, 3 March 1980.

2 Tarapata, G., Paczesny, D., and Tarasiuk, . (2016). Electronic system for floor surface type detection in robotics applications. *International Conference on Optical and Electronic Sensors*, Gdansk, Poland (19–22 June 2016).

3 Santini, F. (2018). Mobile floor-cleaning robot with floor-type detection. US Patent 9,993,129, 12 June 2018.

4 Wang, Y. and Shkel, A.M. (2021). Learning-based floor type identification in ZUPT-aided pedestrian inertial navigation. submitted to *IEEE Sensors Letters* 5 (3): 1–4.

5 Cherkassky, V. and Mulier, F.M. (2007). *Learning from Data: Concepts, Theory, and Methods*. Hoboken, NJ: Wiley.

6 Jolliffe, I.T. and Cadima, J. (2016). Principal component analysis: a review and recent developments. *Philosophical Transactions of the Royal Society A: Mathematical, Physical and Engineering Sciences* 374 (2065): 20150202.

7 Abu-Mostafa, Y.S., Magdon-Ismail, M., and Lin, H.-T. (2012). *Learning from Data*. New York: AMLBook.

8 Chang, C.-B. and Athans, M. (1978). State estimation for discrete systems with switching parameters. *IEEE Transactions on Aerospace and Electronic Systems* 3: 418–425.

9 Hanlon, P.D. and Maybeck, P.S. (2000). Multiple-model adaptive estimation using a residual correlation Kalman filter bank. *IEEE Transactions on Aerospace and Electronic Systems* 36 (2): 393–406.

10 Meng, X., Zhang, Z.-Q., Wu, J.-K. et al. (2013). Self-contained pedestrian tracking during normal walking using an inertial/magnetic sensor module. *IEEE Transactions on Biomedical Engineering* 61 (3): 892–899.

11 Wahlstrom, J., Skog, I., Gustafsson, F. et al. (2019). Zero-velocity detection — a Bayesian approach to adaptive thresholding. *IEEE Sensors Letters* 3 (6): 1–4.

12 Tian, X., Chen, J., Han, Y. et al. (2016). A novel zero velocity interval detection algorithm for self-contained pedestrian navigation system with inertial sensors. *Sensors* 16 (10): 1578.

13 Ma, M., Song, Q., Li, Y., and Zhou, Z. (2017). A zero velocity intervals detection algorithm based on sensor fusion for indoor pedestrian navigation. *IEEE Information Technology, Networking, Electronic and Automation Control Conference (ITNEC)*, Chengdu, China (15–17 December 2017), pp. 418–423.

14 Grimmer, M., Schmidt, K., Duarte, J.E. et al. (2019). Stance and swing detection based on the angular velocity of lower limb segments during walking. *Frontiers in Neurorobotics* 13: 57.

15 Kone, Y., Zhu, N., Renaudin, V., and Ortiz, M. (2020). Machine learning-based zero-velocity detection for inertial pedestrian navigation. *IEEE Sensors Journal*. https://doi.org/10.1109/JSEN.2020.2999863.

16 Wang, Y. and Shkel, A.M. (2019). Adaptive threshold for zero-velocity detector in ZUPT-aided pedestrian inertial navigation. *IEEE Sensors Letters* 3 (11): 1–4.

17 Van Trees, H.L. (2004). *Detection, Estimation, and Modulation Theory, Part I: Detection, Estimation, and Linear Modulation Theory*, 2e. Wiley.

18 Skog, I., Nilsson, J.O., and Händel, P. (2010). Evaluation of zero-velocity detectors for foot-mounted inertial navigation systems. *IEEE International Conference on In Indoor Positioning and Indoor Navigation (IPIN)*, Zurich, Switzerland (15–17 September 2010).

19 Wang, Y., Chernyshoff, A., and Shkel, A.M. (2018). Error analysis of ZUPT-aided pedestrian inertial navigation. *IEEE International Conference on Indoor Positioning and Indoor Navigation (IPIN)*, Nantes, France (24–27 September 2018).

20 Zeitouni, O., Ziv, J., and Merhav, N. (1992). When is the generalized likelihood ratio test optimal? *IEEE Transactions on Information Theory* 38 (5): 1597–1602.

21 Nilsson, J.O., Skog, I., and Händel, P. (2012). A note on the limitations of ZUPTs and the implications on sensor error modeling. *IEEE International Conference on Indoor Positioning and Indoor Navigation (IPIN)*, Sydney, Australia (13–15 November 2012).

22 Wang, Y., Vatanparvar, D., Chernyshoff, A., and Shkel, A.M. (2018). Analytical closed-form estimation of position error on ZUPT-augmented pedestrian inertial navigation. *IEEE Sensors Letters* 2 (4): 1–4.

9

Sensor Fusion Approaches

In previous chapters, we focused on pedestrian inertial navigation with a single Inertial Measurement Unit (IMU). However, IMUs can only measure the specific force and angular rate of the system, while other navigation states, such as the orientation, velocity, and position cannot be directly measured. Many other self-contained sensing modalities can be added to the system to increase the observability of navigation states and thus improve the navigation accuracy. For example, magnetometers can be used to measure the relative orientation between the system and the Earth's magnetic field; barometers can obtain the altitude information by measuring the atmospheric pressure; and ranging techniques can be used to measure relative distances or even position between the transmitter and the receiver. Multiple IMUs can be utilized in a single pedestrian inertial navigation system to improve the navigation results. These approaches fall in the category of sensor fusion technique, where data from multiple sensory sources are combined, so that the output of the system is more accurate than in the cases where these sources are used individually [1]. Extended Kalman Filter is one of the most commonly used method to fuse different sensor readouts. In this chapter, we briefly introduce some of the self-contained sensor fusion approaches that might be used in the pedestrian inertial navigation.

9.1 Magnetometry

Magnetometry is one of the most commonly used techniques, where the position and/or orientation of the system can be derived by measuring the surrounding magnetic field. Compass is a device that detects the direction of the Earth's magnetic field, which is directed approximately toward the North in most part of the surface of the Earth. The compass was invented more than 2000 years ago and was used for navigation in the eleventh century in China [2]. Nowadays, not just the direction but also the amplitude of the Earth's magnetic field can be

Pedestrian Inertial Navigation with Self-Contained Aiding, First Edition. Yusheng Wang and Andrei M. Shkel.
© 2021 The Institute of Electrical and Electronics Engineers, Inc. Published 2021 by John Wiley & Sons, Inc.

measured and used in the navigation of low-earth-orbiting spacecraft (altitude less than 1000 km), where the position of the spacecraft can be estimated with an error less than 10 km only by measuring the magnitude of the Earth's magnetic field [3, 4]. For navigation close to the Earth's surface (altitude less than 30 km), the heading angle can be extracted from the measurement of the Earth's magnetic field vector, if the roll and pitch angles are available from other sources, such as measurement of the gravity. The position information is also available by comparing the measured magnetic field and the World Digital Magnetic Anomaly Map, where the navigation error can reach the same level as GPS [5]. In the case of indoor navigation with complex electromagnetic environment, it has been demonstrated that artificial magnetic field can be created and used to perform navigation [6]. Magnetometry provides directly position and/or orientation information, and therefore eliminates the integration of angular rate and acceleration, which is a major source of navigation errors in inertial navigation. However, it requires a "magnetically clean place" to operate, which limits its applicability [5]. A nonlinear optimization approach was proposed to address the magnetic disturbance due to surrounding iron objects and electric currents in [7]. The root mean heading accuracy could be improved by an order of magnitude by adding magnetometry to inertial navigation.

Magnetometry is one of the most commonly used techniques to be fused with the Zero-Velocity Update (ZUPT)-aided pedestrian inertial navigation. The main reason is that it can provide yaw angle information with high accuracy that is constant over time, whereas the yaw angle estimation error is one of the dominant navigation error sources in the pure ZUPT-aided pedestrian inertial navigation.

9.2 Altimetry

Altimetry is another aiding technique that is widely used for navigation [8]. In this aiding technique, an altimeter is needed to measure the atmospheric air pressure, and the altitude change can be estimated according to the air pressure change. At low altitudes above sea level, the atmospheric pressure decreases approximately linearly as the altitude increases with a rate of about 12 Pa/m. A pressure measurement resolution of 1 Pa, or an altitude measurement accuracy of less than 0.1 m, can be achieved with the currently available commercial micro barometers [9]. Analytical expressions can be derived to correlate the performance of the barometer to the navigation error, showing an altitude estimation accuracy of 1 cm is achievable in a combination of altimetry and ZUPT-aided pedestrian inertial navigation, which is 10× better than the accuracy of the altimetry by itself [10]. Altimetry is simple to implement but is vulnerable to disturbances of environmental pressure and changes in temperature [11].

Another approach to measure the altitude is to use a downward-facing range sensor [12]. Assuming a flat floor surface and a known initial altitude, the altitude of the foot can be estimated by measuring the distance between the foot and the floor by the range sensor. The method has been demonstrated to be able to work when the subject walks upstairs and downstairs, since a discontinuous change in the range sensor readout can be interpreted as the foot moves from one stair to the next. One of the major advantages of this approach over altimeter is that it is more robust to the drift caused by change of atmospheric pressure and temperature. However, walking on the ramp would still be an issue with this approach. It was also reported in [13] that the measurements obtained by the downward-facing range sensor can be utilized to improve the accuracy of the stance phase detection, and thus to improve the overall navigation accuracy of the ZUPT-aided pedestrian inertial navigation.

Like magnetometry, altimetry provides height information with an accuracy independent of total navigation time, and it is also a commonly used technique to be fused with the ZUPT-aided pedestrian inertial navigation to achieve an accurate height estimation during navigation.

9.3 Computer Vision

Another way of implementing estimations of absolute position is computer vision. In this technique, images of the environment are captured and matched against a pre-acquired database for localization [14]. There are four main steps in the computer-vision-based localization: (i) acquire image information; (ii) detect landmarks in current views, such as corners, edges, and objects; (iii) match the detected landmarks with the map; and (iv) update position of the system [15]. This approach potentially requires feature recognition and large databases of the environment. To navigate in a completely unknown environment, the Simultaneous Localization and Mapping (SLAM) method was developed. In this method, as the name indicates, the system simultaneously maps the environment, builds the map, and performs localization. SLAM has been widely applied in self-driving cars, Unmanned Aerial Vehicles (UAVs), and autonomous underwater vehicles. The greatest advantage of SLAM is that it does not require a pre-acquired database of the environment; however, many research challenges are yet to be solved. For example, SLAM is generally computationally intensive, especially in the case of navigation in a large area; it mostly deals with static environment; and it also suffers from odometry drifts [16].

LIght Detection And Ranging (LIDAR) technique shares the same principle of recording and measuring the surrounding environment. In most LIDAR systems, there is a single or multiple laser sources that fire into a rotating mirror to scan the environment. The light is reflected by back-scattering instead of pure reflection and received by the sensor. The distance that the light travels is calculated by multiplying the time difference between transmitting and receiving of the laser by the speed of light. The frequency of the laser used in LIDAR system is typically between infrared and ultraviolet range. For example, light with wavelength of 1064, 532, and 355 nm is typically used. Sampling rate of as high as 1 MHz can be achieved. In this way, a 360° field-of-view can be generated with depth information, which is not directly available in the computer vision aiding. Mapping accuracy of LIDAR can generally reach the order of 0.1 m, and it can be even improved to about 2 cm by acquisition of higher point density of data at the cost of more intensive computation [17]. However, this method does not work well when the air visibility is low, such as in smog and heavy rain. Stereo-vision is currently explored as a method to combine advantages of mono-vision enhanced by ability to provide the depth information. These techniques are also explored to be combined with the infrared vision.

The methods described above obtain information to improve navigation accuracy based on the observation of the environment. Therefore, they are considered by some to be non-self-contained aidings, since their measurements may be interfered or blocked. However, the computer vision aiding in a fully self-contained manner has also been developed, replacing the features in the external environment with the ones within the navigation system. One example is reported in [18], where the camera and the feature to be used are installed on the two feet of a subject. In this way, the relative position between the two feet can be derived as long as the feature is within the field of view of the camera (shown in Figure 9.1). One of the main advantages of this method compared to the other computer vision aiding techniques is that the required computational load is much lower due to the following reasons:

- the shape of the feature to be captured is already known, and it leads to an easier feature recognition;
- instead of a particle filter that is necessary in most SLAM applications, a standard Kalman filter is needed in the algorithm after the relative position information is obtained.

However, the disadvantage is that (i) the measurement in this method is only a relative position, (ii) it cannot completely bound the growth of the position estimation error, and (iii) it required two IMUs mounted on both feet, which increases the system complexity.

Currently, it is not a very common practice to fuse computer vision with the ZUPT-aided pedestrian inertial navigation primarily due to the size limit of the cameras and the high requirement on computational power. However, the unique

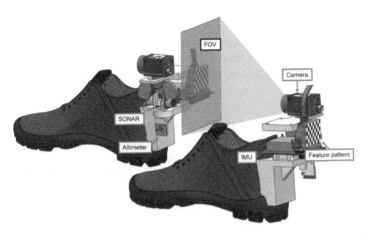

Figure 9.1 Lab-on-Shoe platform. Schematic of the vision-based foot-to-foot relative position measurement. Source: Jao et al. [18].

ability of computer vision to detect the surrounding environment makes it a potential technological candidate.

9.4 Multiple-IMU Approach

Multiple IMUs can be used to work simultaneously in a system to improve the overall performance. The purpose of simultaneously using multiple IMUs mounted at different parts of human body can be categorized into two group: (i) to take advantage of certain features of a specific part of the body, and (ii) to utilize the biomechanical model, i.e. the kinematic relation between different parts of the body.

An example of how we can take advantage of features of a specific part of the body (the first purpose) was reported in [19]. In this method, two IMUs are mounted on the foot and on the head, respectively. The one mounted on the foot is to take advantage of the stance phase during walking and to implement the ZUPT-aided pedestrian inertial navigation. However, the dynamics of the foot during human walking is the severest among all parts of the body, which may cause extra IMU errors due to the shock and vibrations. Therefore, another IMU is mounted on the head to take advantage of the much smoother motion compared to the foot. In this way, the readout from the IMU mounted on the head can be used to calibrate the IMU mounted on the foot. It has also been demonstrated that a much better accuracy in the human activity recognition can be obtained by using two IMUs instead of one.

Another approach is to take advantage of a biomechanical model of the human gait. This approach would typically require multiple IMUs fixed on different parts of human body and subsequently relate the recorded motions of different parts through some known relationships derived from the biomechanical model. For example, a double-pendulum model for the swing phase and an inverted pendulum model for the stance phase may be utilized. Ahmadi et al. [20] reports a complete kinematic model, where seven IMUs are mounted on two feet, two tibias, two thighs, and pelvis, respectively, so that the motion of the entire lower body can be modeled and recorded. In this study, the orientations of the six segments of the lower body are first estimated by gyroscopes. Then, foot-mounted IMUs are used to detect the stance phases of the foot motion. Next, the positions of all segments are estimated after setting the foot segment as root nodes during the stance phases. However, this study is mainly focused on human gait reconstruction, instead of pedestrian inertial navigation. It is also possible to take advantage of some patterns during walking, instead of the full biomechanical model. One simple implementation is to mount two IMUs on two feet, respectively. On top of the two ZUPT-aided inertial navigation on each of the foot, a constraint of maximum allowed separation of the position estimates of the two feet can be imposed [21]. In this way, the systematic errors in position estimates of the two feet will be canceled due to their symmetric nature. In [22], a biomechanical model of walking is used to improve the navigation accuracy, where two IMUs are mounted on upper thigh and foot during navigation. The orientation of the thigh-mounted IMU is used to relate to the kinematic motion of the leg by a biomechanical model. It was demonstrated that by using the approach, the position estimate error can be improved by 50%. Besides adding more information into the system, one of the main advantages of thigh-mounted IMUs and tibia-mounted IMUs over foot-mounted IMUs is that they will experience a much smoother and smaller magnitude of motion, which poses lower requirements on the IMU performance in terms of measurement range, bandwidth, and motion coupling between axes.

9.5 Ranging Techniques

Ranging is a process that measures the relative position between an object and the observer. Different ranging methods include radar, laser, sonar, and LIDAR. For example, radar ranging is the major navigation method for modern aircraft and ship navigation, and sonar ranging is almost the only navigation method for submarine navigation besides inertial navigation. Ranging techniques are generally not considered as self-contained, since it usually involves transmitters and receivers to measure the relative position of the system and some fixed beacons in the environment. However, if the ranging techniques are used to measure the relative position within the system for aiding purposes in inertial navigation, they can be considered as self-contained. In this section, we only focus on the self-contained type of ranging techniques.

9.5.1 Introduction to Ranging Techniques

There are three general categories of ranging techniques currently available: Time-of-Arrival (ToA), Received Signal Strength (RSS), and Angle-of-Arrival (AoA) [23, 24].

9.5.1.1 Time of Arrival
In ToA type of ranging techniques, the system measures the time difference between the signal is transmitted and the signal is received. The distance between the transmitter and receiver can be calculated as the time difference multiplied by the speed at which the signal travels: $\Delta t_1 = t_1^B - t_0^A$, where t_0^A is the transmitting time according to clock A, and t_1^B is the receiving time according to clock B. To precisely determine the time difference, not only very accurate clocks are needed but they also need to be synchronized, which in some cases may not be feasible. Therefore, the so-called two-way ranging is developed to address the issue. In two-way ranging, instead of a transmitter and a receiver, two transceivers (devices comprising both a transmitter and a receiver) are needed. In this configuration, transceiver A transmits a signal at time t_0^A of clock A, and transceiver B receives the signal at time t_1^B of clock B. Next, at time t_2^B of clock B, transceiver B transmits another signal, and the signal is received by transceiver A at time t_3^A of clock A. Then, the ToA of one-way travel of the signal is

$$\Delta t_2 = \frac{(t_3^A - t_0^A) - (t_2^B - t_1^B)}{2}. \tag{9.1}$$

In the two-way ranging configuration, synchronization is not required at the cost of increased complexity of the system. The schematic of the comparison of one-way ranging and two-way ranging is shown in Figure 9.2.

9.5.1.2 Received Signal Strength
RSS type of ranging techniques also measure the distance between the transmitter and the receiver. But, instead of the time difference between signal transmitted and received, the technique utilizes the ratio between the strength of the received signal and the transmitted signal. In condition of unobstructed propagation, the strength of the signal decays at the speed of square of the transmission distance without energy dissipation. Assuming isotropic antenna of both transmitter and

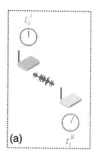 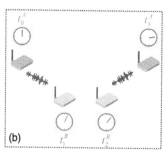

Figure 9.2 Schematic of the comparison of (a) one-way ranging and (b) two-way ranging.

receiver, the Friis equation describes the free space path loss in dB [24]

$$\alpha_{loss} = 20\log_{10}\left(\frac{4\pi d}{\lambda}\right), \tag{9.2}$$

where d is the distance and λ is the wavelength. Notice that Friis equation is only valid when $d \gg \lambda$.

The major advantage of RSS type of ranging is its simplicity of implementation. However, it suffers from possibly unstable transmitter, multipath effect, and dynamic change in the fading channel due to motion of the source and the environment. For example, the signal strength deviations of even only a few dB, which is common in complicated environments, such as indoor environments, can introduce measurement errors as large as 10%.

In indoor pedestrian navigation, WiFi, Bluetooth, and radio-frequency identification (RFID) have been proposed to be used as the ranging source [25–27]. They are all based on RSS type of ranging.

9.5.1.3 Angle of Arrival

AoA type of ranging measures the angle at which the signal arrives at the receiver [28]. A directional antenna or an array of antennas are needed to achieve the goal. The time difference, or the phase difference between the signal received by different receivers in the array is used to estimate the direction of the incoming signal with respect to the array. Note that AoA type of ranging is not just a ranging technique, but also a localization technique, since the distance information combined with the angular information directly yields the position. Unlike the previously mentioned ranging techniques, where at least four pairs of ranging are needed for trilateration in 3D space, theoretically only one AoA ranging pair is enough for localization. However, multipath and shadowing effect may affect the angle measurement and consequently yield large positioning errors.

Typical AoA type of ranging requires a large array of receivers to achieve high accuracy, resulting a relatively large size of the system. Therefore, it is not commonly used in the pedestrian inertial navigation applications [29].

9.5.2 Ultrasonic Ranging

Ultrasonic ranging technique uses ultrasound propagation as the signal to measure the distance. Most ultrasonic ranging techniques are of ToA type. There are two common configurations in ultrasonic ranging implementation: transceiver configuration and transmitter–receiver configuration.

In the transceiver configuration, the transmitter and the receiver are placed at the same place or replaced with a single transceiver to perform ranging. First, the transmitter creates a series of sound pulses, which are often called "pings." The sound pulse is reflected by the object to be measured, and then the receiver detects the reflections (echo) of the pulse. The distance between the system and the object can be calculated with knowledge of the time difference between sending and receiving the pulse and the speed of sound. In this configuration, the system is more compact, and no synchronization is needed. Note that the time difference between two consecutive pulses can be adjusted. A longer time difference results in a higher range of the measurement, and a shorter time difference provides a higher sampling rate of the measurement. However, the pulse of sound might be scattered by the object and as a result, the accuracy of measurement may be low. Figure 9.3 shows the scattering of the sound wave. An ultrasonic wave is created by the transmitter (solid lines) and reflected by an object. The sound wave reaches point B earlier than point A. Therefore, the reflected wave by point B (lighter dashed lines) goes back to the system faster than the reflected wave by point A (darker dashed lines). Since the object is continuous, the duration of the reflected wave may be much longer than the duration when it is created. This causes an error in measurement of the ToA.

In the transmitter–receiver configuration, the transmitter and the receiver are separated. A more accurate point-to-point measurement is available since the scattering effect is eliminated, but synchronization of the clocks associated with the transmitter and the receiver is necessary.

A disadvantage of ultrasonic ranging is that the measurement is not always available. The effect is shown in Figure 9.4. Almost all ultrasonic transmitters are directional, and the created ultrasonic wave is concentrated within an angular cone defined by the properties of the transmitter. If the receiver, or generally the object to be measured is not within the cone region (case (a) in Figure 9.4), no signal will

Figure 9.3 Scattering of the sound wave deteriorates the accuracy of the measurement.

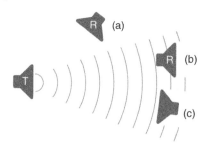

Figure 9.4 "T" stands for transmitter and "R" stands for receiver. Only in case (c) will the receiver receive the signal.

be received. In most case, the receiver also has its sensitive direction. If the ultrasonic wave reaches the receiver from a direction far off its sensitive direction (case (b) in Figure 9.4), still no signal can be received.

9.5.2.1 Foot-to-Foot Ranging

In pedestrian inertial navigation, ultrasonic ranging can be used as a self-contained aiding technique, where the distance between the two feet is measured by ultrasonic ranging sensors. It is also referred as foot-to-foot ultrasonic ranging. In principle, the experimental setup used for foot-to-foot ranging is similar to the one shown in Figure 9.1, except changing the camera and the feature to a transmitter and a receiver.

In such aiding technique, both transceiver configuration and transmitter–receiver configuration can be implemented. Synchronization of clocks of transmitter and receiver can be easily achieved in foot-to-foot ranging by connecting both to the same controller, since they are mounted on the two feet and will not be far away from one another. In such scenario, the transmitter–receiver configuration is preferable to the transceiver configuration to avoid scattering issues and achieve a better measurement accuracy. However, due to a large relative motion between the two feet, the receiver on one foot cannot always stay within the cone region of the transmitter on the other foot. As a result, the ranging measurement is not continuous in most foot-to-foot ranging applications.

A combination of ZUPT aiding and foot-to-foot ranging has been experimentally demonstrated in [30]. In this example, one IMU and two ranging sensors were placed on each foot for pedestrian navigation purpose. Honeywell HG1930 MEMS IMUs were used. Two pairs of ranging sensors are utilized to mitigate the issue of discontinuous ranging measurement. The advantage of foot-to-foot ranging was demonstrated by reducing the error in the yaw angle estimation, and thus reducing the navigation error by 10×.

9.5.2.2 Directional Ranging

Directional ranging is an aiding technique based on ultrasonic ranging sensors. As mentioned in Section 9.5.2, ultrasonic ranging technique is mostly used in ToA type. However, directional ranging has been proposed and demonstrated as an AoA type of ranging for pedestrian inertial navigation in [31].

In pedestrian inertial navigation, the measurement from foot-to-foot ranging technique may not be available during the whole navigation process due to the directivity of the ranging sensors. Directional ranging implementation, instead of being limited by the directivity of the ranging sensor, takes advantage of directionality information to obtain the relative orientation information and improve the overall navigation accuracy. In directional ranging, the ranging measurement not only provides the distance information, its availability indicates that the transmitter and the receiver must be aligned during navigation, enabling the directional ranging to obtain the relative orientation information between the two feet. The ability of directional ranging to measure the relative angle makes it a good complement of ZUPT-aided navigation algorithm by granting the system partial observability of yaw angle, which is the dominant error source in navigation with only ZUPT aiding.

In the reported experimental setup (Figure 9.5a), the transmitter–receiver configuration is implemented, with the transmitter and the receiver separated and mounted on two feet. The output of the ranging system is zero if the transmitter and the receiver are not aligned. The major advantage of the setup in Figure 9.5a is the ability to measure the distance between the transmitter and the receiver and a relative orientation between the two feet. When the transmitter and the receiver are aligned, Figure 9.5b, the ranging system will obtain the distance information. The full alignment requires two conditions: (i) two feet are aligned along the direction of transmission of the ultrasonic wave (a counter-example is shown in Figure 9.5c), and (ii) yaw angles are the same for both feet (a counter-example is shown in Figure 9.5d). The two conditions can be mathematically expressed as

$$\arctan \frac{N_l - N_r}{E_l - E_r} - \text{Yaw}_r = \pm 75°, \tag{9.3}$$

$$\text{Yaw}_r - \text{Yaw}_l = 0, \tag{9.4}$$

where Yaw, N, and E correspond to the yaw angle, and positions along the North and the East, respectively. The subscripts l and r indicate the left foot and the right foot.

Experiments were conducted both indoors and outdoors to verify the effects of directional ranging, and in both cases, only the self-contained navigation was used. The indoor test had a total navigation time of about three minutes. The results with different navigation algorithms are shown in Figure 9.6. The dashed line is the ground truth based on the floor plan, and the solid lines in darker and lighter gray are the estimated trajectories of the left and the right foot, respectively. The estimation drifted away fast when no compensation mechanism was implemented. ZUPT aiding during the navigation suppressed the navigation error, but the yaw

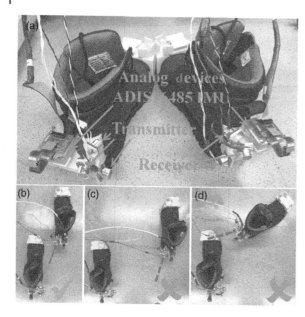

Figure 9.5 (a) Experimental setup of the illustrative experiment; (b) Ranging data are collected with transmitter and receiver aligned; (c) and (d) Ranging data are not collected with transmitter and receiver not aligned. Dashed lines in (b)–(d) are directions of transmission of the ultrasonic wave. Source: Wang et al. [31].

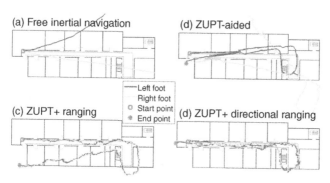

Figure 9.6 A comparison of results of different aiding techniques for indoor environment. Source: Wang et al. [31].

angle error was not compensated, as shown in Figure 9.6b. Foot-to-foot ranging brought the two estimated trajectories for the two feet together, and also partially compensated for the yaw angle error, Figure 9.6c, while the directional ranging further improved the estimation and reduced the overall navigation errors by 1.8 times as compared with the regular foot-to-foot ranging, Figure 9.6d.

Outdoor navigation experiment was conducted with the navigation time of about six minutes and the total navigation length of around 420 m. The estimated trajectories obtained by different ranging techniques are presented in Figure 9.7.

Figure 9.7 A comparison of different aiding techniques for self-contained navigation. The dashed line is the ground truth. The estimated ending points are denoted by the dots. The total navigation length was around 420 m. Source: Wang et al. [31].

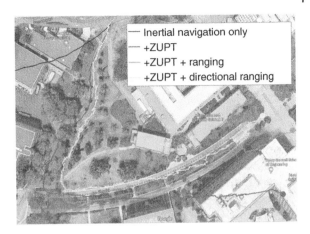

Estimation error of position was reduced from 25 to 10 m by implementing directional ranging instead of foot-to-foot ranging.

In the illustrated directional ranging implementation, the configuration is exactly the same as foot-to-foot ranging. No increase in complexity of the hardware is required as compared to the foot-to-foot ranging implementation, and the navigation accuracy improvement is achieved purely algorithmically.

9.5.3 Ultrawide Band Ranging

Ultrawide Band (UWB) is defined as a wireless system with absolute bandwidth of greater than 500 MHz or relative bandwidth larger than 20% [32]. Such a wide bandwidth increases the reliability of the ranging system, since the different frequency components increase the probability of finding available direct path signal. ToA type of ranging is most popular with the UWB techniques due to its ability to achieve a centimeter-level measurement accuracy. For a single-path additive white Gaussian noise channel, the best achievable accuracy of a distance estimation according to the Cramer–Rao Lower Bound using ToA techniques is [33]:

$$\sigma_{\hat{d}} = \frac{c}{2\sqrt{2}\pi \text{BW}_e \sqrt{\text{SNR}}}, \tag{9.5}$$

where c is the speed of light, BW_e is the effective bandwidth, and SNR is signal-to-noise ratio. Equation (9.5) illustrates the advantage of UWB that a larger bandwidth of UWB signal would result in smaller measurement uncertainty, i.e. higher measurement accuracy.

Similar to ultrasonic ranging techniques, UWB ranging can be used in a self-contained arrangement for pedestrian inertial navigation. Due to a relatively larger measurement range of UWB propagation compared to ultrasonic ranging, the UWB ranging can be used not only in foot-to-foot ranging but also in agent-to-agent ranging. This ranging is also used in so-called cooperative localization. If we consider the network of multiple agents as a whole system, cooperative localization can be treated as self-contained since no preinstalled anchor (or beacon) is involved in the network [34].

Location estimation with UWB ranging is straightforward in a line-of-sight (LOS) and single-path environment. However, in a more realistic situation, such as indoor environment, where direct LOS might be blocked due to cubicles, doors, hallways, etc., measurement errors will be inevitable. For example, multipath propagation results in the overlap of multiple replicas of the transmitter signal and induces range measurement errors. Non-line-of-sight (NLOS) propagation will result in the reflection of the signal. Therefore, the signal travels an extra distance before it is received, and the corresponding positive bias is called the NLOS error. Further details on UWB ranging can be found in [35].

The effects of interagent cooperative localization by UWB ranging have been reported in literature. For example, in [36], ZUPT-aided pedestrian inertial navigation was combined with UWB interagent ranging, and it was demonstrated that the Root Mean Square Error (RMSE) of the navigation could be improved as the square root of the number of agents. In [37], it has been demonstrated that cooperative localization by UWB in a three-person team might reduce the position errors by up to 70% with consumer grade IMUs.

9.6 Conclusions

Due to the limits of inertial navigation, some navigation states (position, velocity, and orientation) are unobservable with only IMUs. Therefore, many other sensing modalities can be utilized to increase the observability of the navigation system. In this chapter, we only introduced some of the most commonly used techniques that can be implemented in a self-contained paradigm: magnetometry, altimetry, computer vision, multiple-IMU, and ranging techniques.

All these techniques have their unique application scenarios, and a proper combination of them would be desirable to maximize the precision, robustness, and autonomy of the overall system. Such a combination has not been achieved yet, and we provide our perspective on further development of pedestrian inertial navigation systems in the Chapter 10.

References

1 Kam, M., Zhu, X., and Kalata, P. (1997). Sensor fusion for mobile robot navigation. *Proceedings of the IEEE* 85 (1): 108–119.

2 Merrill, R.T. and McElhinny, M.W. (1983). *The Earth's Magnetic Field: Its History, Origin and Planetary Perspective*, 2nd printing ed. San Francisco, CA: Academic press.

3 Psiaki, M.L., Huang, L., and Fox, S.M. (1993). Ground tests of magnetometer-based autonomous navigation (MAGNAV) for low-earth-orbiting spacecraft. *Journal of Guidance, Control, and Dynamics* 16 (1): 206–214.

4 Shorshi, G. and Bar-Itzhack, I.Y. (1995). Satellite autonomous navigation based on magnetic field measurements. *Journal of Guidance, Control, and Dynamics* 18 (4): 843–850.

5 Goldenberg, F. (2006). Geomagnetic navigation beyond the magnetic compass. *2006 IEEE/ION Position, Location, and Navigation Symposium*, Coronado, CA, USA (25–27 April 2006).

6 Hellmers, H., Norrdine, A., Blankenbach, J., and Eichhorn, A. (2013). An IMU/magnetometer-based indoor positioning system using Kalman filtering. *International Conference on Indoor Positioning and Indoor Navigation (IPIN)*, Montbeliard-Belfort, France (28–31 October 2013).

7 Wu, J. (2019). Real-time magnetometer disturbance estimation via online nonlinear programming. *IEEE Sensors Journal* 19 (12): 4405–4411.

8 Romanovas, K., Goridko, V., Klingbeil, L. et al. (2013). Pedestrian indoor localization using foot mounted inertial sensors in combination with a magnetometer, a barometer and RFID. In: *Progress in Location-Based Services*, 151–172. Berlin, Heidelberg: Springer-Verlag.

9 TDK InvenSense (2019). ICP-10100 Barometric Pressure Sensor Datasheet. https://invensense.tdk.com/wp-content/uploads/2018/01/DS-000186-ICP-101xx-v1.2.pdf.

10 Jao, C.-S., Wang, Y., Askari, S., and Shkel, A.M. (2020). A closed-form analytical estimation of vertical displacement error in pedestrian navigation. *IEEE/ION Position, Location and Navigation Symposium (PLANS)*, Portland, OR, USA (20–23 April 2020).

11 Parviainen, J., Kantola, J., and Collin, J. (2008). Differential Barometry in personal navigation. *IEEE/ION Position, Location and Navigation Symposium (PLANS)*, Monterey, CA, USA (5–8 May 2008).

12 Jao, C.-S., Wang, Y., and Shkel, A.M. (2020). A hybrid barometric/ultrasonic altimeter for aiding ZUPT-based inertial pedestrian navigation systems. *ION GNSS+ 2020*, 21–25 September 2020.

13 Jao, C.-S., Wang, Y., and Shkel, A.M. (2020). A zero velocity detector for foot-mounted inertial navigation systems aided by downward-facing range sensor. *IEEE Sensors Conference 2020*, 25–28 October 2020.

14 Kourogi, M. and Kurata, T. (2003). Personal positioning based on walking locomotion analysis with self-contained sensors and a wearable camera.

IEEE/ACM International Symposium on Mixed and Augmented Reality, Tokyo, Japan (7–10 October 2003).

15 Bonin-Font, F., Ortiz, A., and Oliver, G. (2008). Visual navigation for mobile robots: a survey. *Journal of Intelligent and Robotic Systems* 53 (3): 263–296.

16 Thrun, S. (2007). Simultaneous localization and mapping. In: (ed. Bruno Siciliano, Oussama Khatib, Frans Groen) *Robotics and Cognitive Approaches to Spatial Mapping*, 13–41. Berlin, Heidelberg: Springer-Verlag.

17 Amzajerdian, F., Pierrottet, D., Petway, L. et al. (2011). Lidar systems for precision navigation and safe landing on planetary bodies. *International Symposium on Photoelectronic Detection and Imaging 2011: Laser Sensing and Imaging; and Biological and Medical Applications of Photonics Sensing and Imaging*, Volume 8192, International Society for Optics and Photonics, p. 819202.

18 Jao, C.-S., Wang, Y., and Shkel, A.M. (2020). Pedestrian inertial navigation system augmented by vision-based foot-to-foot relative position measurements. *IEEE/ION Position, Location and Navigation Symposium (PLANS)*, Portland, OR, USA (20–23 April 2020).

19 Askari, S., Jao, C.-S., Wang, Y., and Shkel, A.M. (2019). Learning-based calibration decision system for bio-inertial motion application. *IEEE Sensors Conference*, Montreal, Canada (27–30 October 2019).

20 Ahmadi, A., Destelle, F., Unzueta, L. et al. (2016). 3D human gait reconstruction and monitoring using body-worn inertial sensors and kinematic modeling. *IEEE Sensors Journal* 16 (24): 8823–8831.

21 Skog, I., Nilsson, J.O., Zachariah, D., and Handel, P. (2012). Fusing the information from two navigation systems using an upper bound on their maximum spatial separation. *IEEE International Conference on Indoor Positioning and Indoor Navigation (IPIN)*, Sydney, Australia (13–15 November 2012).

22 Ahmed, D.B. and Metzger, K. (2018). Wearable-based pedestrian inertial navigation with constraints based on biomechanical models. *IEEE/ION Position, Location and Navigation Symposium (PLANS)*, Monterey, CA, USA (23–26 April 2018).

23 Farahani, S. (2011). *ZigBee Wireless Networks and Transceivers*. Newnes.

24 Frattasi, S. and Della Rosa, F. (2017). *Mobile Positioning and Tracking: From Conventional to Cooperative Techniques*. Wiley.

25 Cheng, J., Yang, L., Li, Y., and Zhang, W. (2014). Seamless outdoor/indoor navigation with WIFI/GPS aided low cost Inertial Navigation System. *Physical Communication* 13: 31–43.

26 Nguyen, K. and Luo, Z. (2013). Evaluation of bluetooth properties for indoor localisation. In: *Progress in Location-Based Services*, 127–149. Berlin, Heidelberg: Springer-Verlag.

27 Ruiz, A.D.J., Granja, F.S., Honorato, J.C.P., and Rosas, J.I.G. (2012). Accurate pedestrian indoor navigation by tightly coupling foot-mounted IMU and RFID measurements. *IEEE Transactions on Instrumentation and Measurement* 61 (1): 178–189.

28 Dotlic, I., Connell, A., Ma, H. et al. (2017). Angle of arrival estimation using decawave DW1000 integrated circuits. *IEEE Workshop on Positioning, Navigation and Communications (WPNC)*, Bremen, Germany (25–26 October 2017).

29 Wielandt, S. and De Strycker, L. (2017). Indoor multipath assisted angle of arrival localization. *Sensors* 17 (11): 2522.

30 Laverne, M., George, M., Lord, D. et al. (2011). Experimental validation of foot to foot range measurements in pedestrian tracking. *ION GNSS Conference*, Portland, OR, USA (19–23 September 2011).

31 Wang, Y., Askari, S., Jao, C.S., and Shkel, A.M. (2019). Directional ranging for enhanced performance of aided pedestrian inertial navigation. *IEEE International Symposium on Inertial Sensors & Systems*, Naples, FL, USA (1–5 April 2019).

32 FCC, Office of Engineering and Technology (2002). *Revision of Part 15 of the Commission's Rules Regarding Ultra-Wideband Transmission Systems. ET Docket, no. 98-153.*

33 Gezici, S., Tian, Z., Giannakis, G.B. et al. (2005). Localization via ultra-wideband radios: a look at positioning aspects for future sensor networks. *IEEE Signal Processing Magazine* 22 (4): 70–84.

34 Wymeersch, H., Lien, J., and Win, M.Z. (2009). Cooperative localization in wireless networks. *Proceedings of the IEEE* 97 (2): 427–450.

35 Oppermann, I., Hamalainen, M., and Iinatti, J. (2005). *UWB: Theory and Applications*. Wiley.

36 Nilsson, J.O., Zachariah, D., Skog, I., and Handel, P. (2013). Cooperative localization by dual foot-mounted inertial sensors and inter-agent ranging. *EURASIP Journal on Advances in Signal Processing* (1): 164.

37 Olsson, F., Rantakokko, J., and Nygards, J. (2014). Cooperative localization using a foot-mounted inertial navigation system and ultrawideband ranging. *IEEE International Conference on Indoor Positioning and Indoor Navigation (IPIN)*, Busan, Korea (27–30 October 2014).

10

Perspective on Pedestrian Inertial Navigation Systems

Pedestrian inertial navigation has been of high interest in many fields, such as human health monitoring, personal indoor navigation, and localization systems for first responders. In this chapter, we provide a perspective on further development of both the hardware and the software for pedestrian navigation.

10.1 Hardware Development

Hardware development for the pedestrian inertial navigation mainly aims to solve the problem of incorporating different sensing modalities with a reasonable size and weight, such that the overall system is compact, robust, and accurate.

Due to complex environment in which a pedestrian may walk, external reference signals may not always be available. Therefore, a self-contained navigation system is preferable to guarantee the functionality and robustness of the system in different scenarios. Available aiding techniques include the ZUPT-aided pedestrian navigation algorithm, foot-to-foot ranging, etc., which have been discussed in previous chapters.

On the other hand, due to the nature of self-contained navigation, although the navigation error can be limited by the aiding techniques, they cannot be bounded within a fixed range as navigation time increases. As a result, we believe that information about the environment should be utilized whenever it is available to improve navigation accuracy. Signal of Opportunity (SoP) is an emerging approach for aiding in navigation [1]. Generally speaking, SoP refers to the use of any signals for navigation, which are not normally intended for navigation. For example, cellular signals (CDMA and LTE) for outdoor environment and WiFi for indoor environment can be utilized due to their prevalence.

Furthermore, cooperative localization is also desirable for a team of mobile agents, with communication and compensation capabilities. Jointly processing a relative measurement between any two agents will increase localization accuracy.

Pedestrian Inertial Navigation with Self-Contained Aiding, First Edition. Yusheng Wang and Andrei M. Shkel.
© 2021 The Institute of Electrical and Electronics Engineers, Inc. Published 2021 by John Wiley & Sons, Inc.

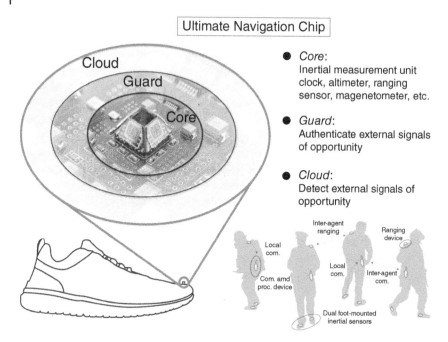

Figure 10.1 Our perspective of pedestrian inertial navigation system: uNavChip [2, 3].

In our view, the pedestrian inertial navigation system of the future, or Ultimate Navigation Chip (uNavChip), should be based on simultaneous integration of deterministic, probabilistic, and cooperative localization information. The concept of uNavChip is shown in Figure 10.1. The core of the system is the deterministic localization which can be achieved by Micro-Electro-Mechanical System (MEMS) Inertial Measurement Unit (IMU) and clock for inertial navigation and timing, as well as Capacitive Micro-machined Ultrasonic Transducer (CMUT) for foot-to-foot proximity ranging, altimeter for altitude measurement, camera for computer vision, and magnetic sensor for relative orientation measurement. The localization unit is envisioned to be integrated in sole of shoes, for the most efficient compensation of drift. The IMU will derive the absolute orientation and position, which would be regularly re-calibrated by zero velocity updates with other sensor inputs. The ultimate vision is to have the IMU and all other aiding techniques co-fabricated in parallel on both sides of a single silicon substrate and subsequently folded into a cube and locked in place using specially designed micro-lock mechanisms [2]. On top of deterministic localization, the probabilistic localization approach aims to exploit a cloud of signals of opportunity, which are abundant but could be usable in GPS challenged environments, turning cellular transmitters to their own "dedicated pseudolites" [3]. Another critical aspect is to detect and resolve spoofing and jamming of signals of opportunity, i.e. to provide an authentication of SoP. Algorithms are needed to determine how much signals of opportunity can be trusted and utilized for navigation augmentation. Cooperative localization can be achieved if multiple mobile agents are available in the network, with communication and computation capabilities, jointly processing a relative measurement between each agent to increase their localization accuracy. An integrated collaborative positioning framework that will utilize synthetic aperture navigation will be needed, aiding an inertial navigation system in an ultra-tightly coupled fashion with cellular signals.

Most of the technologies mentioned above exist and are possible to implement today. If we simply take off-the-shelf components and integrate them into a single system, all functionalities can be achieved with the system of the size of a watermelon. However, it is not realistic to perform navigation in the field with such a large system. Therefore, miniaturization is needed to reduce the size of the system. With some of the available technologies, it is possible to reduce the size of the system to the size of an apple. We envision that the size of uNavChip can be reduced to the size of an apple seed, when ultrahigh density, heterogeneous integration, and MEMS integration techniques are realized simultaneously. Such high density integration will provide maximum autonomy, security, and precision.

10.2 Software Development

Software development for the pedestrian inertial navigation mainly aims to explore algorithms to fully use the collected data, in order to further improve the navigation accuracy and adaptivity without too much computational load.

For example, there are many ways to fully utilize the IMU data besides just integrating them into position estimation. Stance phase detection, floor type detection, and gait frequency extraction are some other possible ways to use the data, as has been discussed in previous chapters. Data obtained by other sensing modalities can be used in a similar manner. Machine learning approach is probably a good candidate to achieve the full use of the obtained data. For example, whenever GPS and LTE signals are available, they can not only be used for localization but the position information can also be compared against the map to obtain the floor type. Such information can be used to train the system, so that the floor type detection can be more accurate even when the GPS signals are not available.

10.3 Conclusions

To conclude, pedestrian inertial navigation has drawn a lot of interest due to its wide application and availability of various types of supporting technologies. To make pedestrian inertial navigation more accurate, adaptive, and robust, several approaches can be taken:

- introduction of extra sensing modalities by sensor fusion;
- improvement of the performance of every sensor used in the system by better sensor designs; and
- full utilization of the data obtained by all sensors by algorithms such as machine learning.

In order to achieve the goal from all three approaches, further development of MEMS technologies is necessary for various sensors with higher performance and smaller size and weight. Besides, innovative algorithms are also needed to process the data.

References

1 Morales, J., Roysdon, P., and Kassas, Z. (2016). Signals of opportunity aided inertial navigation. *ION GNSS Conference*, Portland, Oregon (12–16 September 2016).

2 Shkel, A.M., Kassas, Z., and Kia, S. (2019). uNavChip: Chip-Scale Personal Navigation System Integrating Deterministic Localization and Probabilistic Signals of Opportunity. Public Safety Innovation Accelerator Program, US Department of Commerce, National Institute of Standards and Technology (NIST), 70NANB17H192.

3 Nilsson, J.-O., Zachariah, D., Skog, I., and Handel, P. (2013). Cooperative localization by dual foot-mounted inertial sensors and inter-agent ranging. *EURASIP Journal on Advances in Signal Processing* (1): 164.

4 Efimovskaya, A., Lin, Y.-W., and Shkel, A.M. (2017). Origami-like 3-D folded MEMS approach for miniature inertial measurement unit. *IEEE/ASME Journal of Microelectromechanical Systems* 26 (5): 1030–1039.

5 Shamaei, K., Khalife, J., and Kassas, Z.M. (2018). Exploiting LTE signals for navigation: theory to implementation. *IEEE Transactions on Wireless Communications* 17 (4): 2173–2189.

Index

Pedestrian Inertial Navigation with Self-Contained Aiding, First Edition. Yusheng Wang and Andrei M. Shkel.
© 2021 The Institute of Electrical and Electronics Engineers, Inc. Published 2021 by John Wiley & Sons, Inc.

IEEE Press Series on Sensors

Series Editor: Vladimir Lumelsky, Professor Emeritus, Mechanical Engineering, University of Wisconsin-Madison

Sensing phenomena and sensing technology is perhaps the most common thread that connects just about all areas of technology, as well as technology with medical and biological sciences. Until the year 2000, IEEE had no journal or transactions or a society or council devoted to the topic of sensors. It is thus no surprise that the IEEE Sensors Journal launched by the newly-minted IEEE Sensors Council in 2000 (with this Series Editor as founding Editor-in-Chief) turned out to be so successful, both in quantity (from 460 to 10,000 pages a year in the span 2001–2016) and quality (today one of the very top in the field). The very existence of the Journal, its owner, IEEE Sensors Council, and its flagship IEEE SENSORS Conference, have stimulated research efforts in the sensing field around the world. The same philosophy that made this happen is brought to bear with the book series.

Magnetic Sensors for Biomedical Applications
Hadi Heidari, Vahid Nabaei

Smart Sensors for Environmental and Medical Applications
Hamida Hallil, Hadi Heidari

Whole-Angle MEMS Gyroscopes: Challenges, and Opportunities
Doruk Senkal and Andrei M. Shkel

Optical Fibre Sensors: Fundamentals for Development of Optimized Devices
Ignacio Del Villar and Ignacio R. Matias.

Pedestrian Inertial Navigation with Self-Contained Aiding
Yusheng Wang and Andrei M. Shkel

Printed and bound by CPI Group (UK) Ltd, Croydon, CR0 4YY
04/10/2021

03085484-0001